THE COMPANY BUSINESS

BOOK THREE - THE CIA AREA 51 CHRONICLES
1955–1979

The Complete Illustrated History of the CIA at Area 51

By

TD Barnes
"Thunder"

TD Barnes Copyright 2018

TABLE OF CONTENTS

Dedication

I dedicate this book about the Soviet MiG exploitation projects in Nevada to all who encountered in combat the Mikoyan-Gurevich, commonly known as the MiG. I also dedicate it to the Area 51 veterans whose contribution at the beginning made Area 51 what it is today.

Acknowledgements

I acknowledge all of you with whom I served at the CIA's Area 51. Included are all those named and those whose names I cannot reveal for national security reasons. I thank you for your service. Your participation saved the lives of many aircrews then, now, and into the future. I acknowledge Col Gail Peck and all the bandits of the Red Eagles and Project CONSTANT PEG who flew the MiGs on the Tonopah Test Range. I recognize the Red Hats, whose legacy is still classified. You bandits and Red Hats carried forward what the CIA, the Air Force, the Navy, and my special projects team learned while exploiting the MiGs at Area 51. This combined effort has saved countless lives of our brothers and sisters in combat ever since and will continue to do so in future wars. I also acknowledge my friends Pete Merlin, Joerg Arnu, and Alan Johnson for supporting and carrying on the legacy of the Area 51 veterans.

The Agencies Participating in the MiG Exploitation Projects at Area 51

Declassification:

The National Air Intelligence Center declassified the formerly Top Secret HAVE DOUGHNUT, HAVE DRILL, and HAVE FERRY FTD reports referenced in this book in September 1997. Reviewed by NAIC/PA and NAIC/TAA to comply with IAW Executive Order 12958. FTD, TAC, Navy, AFFTC, SAC, ADC, and ASD for DIA referenced as HAVE DOUGHNUT and HAVE DRILL/HAVE FERRY: FTD-CR-20–13–69-INT Vol. I, "HAVE DOUGHNUT Technical," 1 November 1968, FTD-CR-20–02–69 Vol. I, "HAVE DRILL, HAVE FERRY Technical," 1 November 1969, and FTD-CR-20–02–69 Vol. H, "HAVE DRILL/HAVE FERRY Tactical," 1 April 1970.

On August 16, 2013, the CIA acknowledged its role in the MiG exploitation projects with the release of "The Area 51 File" titled by the National Security Archive as, "The CIA Declassifies Area 51." Secret Aircraft and Soviet MiG Declassified Documents Describe Stealth Facility in Nevada. The Security Archive Electronic Briefing Book #443 further describes the release. Following this declassification, the author presented this book to the CIA PBR for review March 10, 2014. The author received credit in the

CIA document declassification of August 16, 2013.

https://nsarchive.wordpress.com/2013/08/16/the-cia-declassifies-area-51/

The declassification of the Air Forces' role in the exploitation at Area 51 occurred during the 2006 declassification of the CONSTANT PEG program, where the USAF held a series of press conferences about the former top-secret US MiG programs in Nevada. Declassification revealed that in addition to the classified exploitation flights at Groom Lake starting in 1968, the US flew the Soviet-built MiG more than 15,000 sorties outside of Area 51. The US trained approximately 7,000 aircrews against dissimilar MiG aggressors in the Nevada desert between 1980 and the end of the program in 1988.

Preface

In 1954, when Gen Curtis LeMay's Air Force refused to do it, President Eisenhower tasked the Central Intelligence Agency with developing a high-flying reconnaissance plane to overfly the Soviet Union. The decision sparked months of political posturing, pitting the CIA vs. the US Air Force, the Bell Aircraft, Martin Aircraft, and Fairchild Engine and Airplane vs. the Lockheed Aircraft Company. Lockheed in Burbank, California won the bid to build such a plane that became known as the U-2. Lockheed called it the Angel because it flew so high. The Air Force called it the Dragon Lady because the aircraft was so unforgiving. The CIA called it the U-2, code-named the Project AQUATONE. The birthplace of this marvelous new aerial concept was known as the Skunk Works. The CIA's flight test center was called Area 51.

The genesis of Area 51 was the CIA and its contractor, Lockheed Aircraft Co. needing a secure test site for its U-2. Area 51 was selected when Lockheed's U-2 designer Clarence L. "Kelly" Johnson sent project pilot Tony LeVier and Lockheed Skunk Works chief foreman Dorsey Kammerer on a two-week survey mission to scout locations for a new flight test facility in an unmarked Beechcraft V-35 Bonanza.

Richard M. Bissell, Jr., "special assistant" to CIA director Allen Dulles, and director of the AQUATONE program reviewed fifty potential sites with his Air Force liaison, Col. Osmond J. "Ozzie" Ritland. None of the sites seemed to meet the stringent security requirements of the program. They rejected Johnson's proposed Mud Lake site near Tonopah, Nevada because it was too close to populated areas. Ritland, however, recalled "a little X-shaped field" just off the eastern side of Groom Dry Lake, about 100 miles north of Las Vegas, Nevada, just outside the Atomic Energy Commission's (AEC) nuclear proving ground at Yucca Flat.

In April 1955, Tony LeVier, Kelly Johnson, Dick Bissell, and Col. Osmond Ritland flew out to Nevada on a two-day survey of the most promising lakebeds, including Groom Lake. The abandoned airfield that Ritland had remembered was sandy, overgrown, and unusable, but the three-mile-wide dry lakebed was perfect.

Bissell secured a presidential action adding the Groom Lake area to the AEC proving ground. Ritland wrote three memos to the Air Force, AEC, and the Training Command that administered the gunnery range. Signed by Assistant Air Secretary for Research and Development Trevor Gardner, they ensured that range activities would not impinge on the new test site. Security for the project was now assured.

Johnson met with CIA officials in Washington, DC and discussed progress on the flight test facility and the AQUATONE program. His proposal to name the flight test facility "Paradise Ranch" was accepted. It was an ironic choice which, he later admitted was "a dirty trick to lure workers to the program." However, the CIA chose the name "Watertown," the hometown of the CIA boss, DCI Dulles. Nonetheless, Johnson's Paradise Ranch name remained and would dominate the name game when the CIA resumed operations for the follow-on Project OXCART.

In May 1955, LeVier, Kammerer, and Johnson returned to Groom Lake in Lockheed's Bonanza. Using a compass and surveying equipment, they laid out a place for a 5,000-foot, north-south runway on the southwest corner of the lakebed. They also staked out the general layout of the base. Herb Miller of CIA Development Projects Staff used the Atomic Energy Commission (AEC) as a cover to issue $800,000 in contracts for construction of the flight test facility and organize a team of construction crews.

An AEC information booklet called Background Information on Nevada Nuclear Tests published two years later, in 1957, gave a cover story for the Watertown operation. It stated that the National Advisory Committee for Aeronautics (NACA) was operating U-2 aircraft at the Groom Lake site with logistical and technical support from the Air Weather Service of the US Air Force to make weather observations at heights unattainable by most aircraft. At that time, the aircraft were unpainted except for fictitious NACA markings to provide a cover if the agency lost one of them over denied territory.

Seth Woodruff, Jr., Manager of the AEC Las Vegas Field Office, announced to the news media that he had "instructed the Reynolds Electrical and Engineering Co., Inc. [REECo] to begin preliminary work on a small, satellite Nevada Test Site installation." He noted that work was already underway at the location "a few miles northeast of Yucca Flat and within the Las Vegas Bombing and Gunnery Range." Woodruff said that the installation would include "a runway, dormitories, and a few other buildings for housing equipment." He described the CIA station at Area 51 as an "essentially temporary" facility. The press release distribution went to 18 media outlets in Nevada and Utah including a dozen newspapers, four radio stations, and two television stations.

LeVier and fellow Lockheed test pilot Bob Matye spent nearly a month at Groom Lake removing surface debris left from gunnery practice during World War II. LeVier also drew up a proposal for four three-mile-long runways to be marked on the hard-packed clay. Johnson, however, refused to approve the $450.00 expense, citing a lack of funds. Drilling resulted in the discovery of a limited water supply, but trouble with the well soon developed, which required trucking in water until subsequent drilling successfully located water for the facility.

In July 1955, the completed construction of the flight test facility consisted of a single paved 5,000-foot runway, three hangars, a control tower, and rudimentary accommodations for test personnel. The base's few amenities included a movie theater and volleyball court. Additionally, there was a mess hall, and several water wells and fuel storage tanks. CIA, Air Force, and Lockheed personnel began arriving at Watertown with CIA-assigned Richard Newton as flight test facility commander. Thus, the CIA's Area 51 came to be.

A Preview into What This Meant to the CIA's Future.

In June 1957, CIA pilot classes finished training. The U-2 test operation moved to North Base at Edwards AFB, California. The Air Force assigned its operational U-2 aircraft to the 4028th Strategic Reconnaissance Squadron at Laughlin, Texas, leaving Watertown a virtual ghost town. The flight test facility went into caretaker status with a site manager, security, and minimal complement of personnel present.

In September 1959, the CIA's Area 51 station gained new life with EG&G agreed to move its radar test facility to Groom Lake. The company constructed a special pylon on a paved loop road on the western side of the lakebed for radar cross-section evaluations of the CIA's latest plane, the high-flying, Mach 3 replacement for the U-2.

A year later, in September 1960, flight test facility construction began at Area 51 to build facilities to support Project OXCART, the Lockheed A-12. The existing 5,000-foot runway designed for the U-2 was incapable of supporting the weight of the A-12, requiring the construction of a new airstrip (Runway 14/32) that in November 1960, gave the A-12 its required runway of at least 8,500 feet long and 150-feet-wide. A 10,000-foot hard asphalt extension, with a concrete turnaround pad in the middle, cut diagonally across the southwest corner of the lakebed.

In 1962, the Central Intelligence Agency would add the Science and Technology Doctorate to continue what it had started at Area 51.

During the next eight years, the CIA would build and test America's first stealth plane, the A-12, the fastest and highest-flying air-breathing aircraft ever built, a record the CIA still holds today. Area 51 provided three A-12s and rotated the needed personnel to Kadena, Okinawa for Operation BLACK SHIELD, overflights of North Vietnam, Laos, Cambodia, and North Korea. This time, the CIA did not close its station as the Project OXCART ended with the Air Force SR-71s replacing the CIA's A-12s.

While the Project OXCART people prepared to leave, the CIA and its EG&G special projects cadre prepared to assume a new role with a smaller team and multiple customers. The CIA saw that the United States needed to look to the future, to the day that the nation's security needed more than the high-flying, Mach 3 surveillance platform. The CIA saw the need to continue with the stealth efforts first realized with the U-2 and embedded in the A-12. The agency now had the new science and technology division whose business was running a new type of station called Area 51, a secret venue with the specialized equipment, and the special projects specialists to expand its science and technology into the future.

The agency was no longer just for intelligence gathering. It was now in the business of science and technology, a business whose first customer was the US Air Force and Navy seeking answers to why they were losing the air war in Vietnam. In January 1968, as Project OXCART moved out, the CIA welcomed in Project HAVE DOUGHNUT, a joint USAF/Navy technical and tactical evaluation of the MiG-21F-13 to determine why it was enjoying a 9:1 kill ratio against the United States Air Force, Navy, and Army in the undeclared Vietnam war. Gone five months later were the CIA and Lockheed pilots identified by a Dutch number, the A-12 Articles replaced by Soviet MiGs flown by the Navy, and Air Force pilots called bandits.

Chapter 1 -A Black World Business

When John McCone, an engineer, and manager of large corporations, became the director of Central Intelligence in November 1961, he made it his top priority to establish a new directorate that would consolidate the Agency's far-flung and uncoordinated science and technology research, collection, and analysis. His first major reorganization and redirection were to promote efficiency, productivity, and innovation by the formation of the Deputy Directorate of Research (DDR) on February 19, 1962, under Herbert "Pete" Scoville. Unfortunately, Scoville failed to form the robust directorate that McCone wanted and submitted his resignation on April 25, 1963.

At this point, DCI McCone replaced Scoville with 34-year-old Albert Dewell "Bud" Wheelon—the Directorate of Intelligence's Assistant Deputy for Scientific Intelligence and on August 5, 1963. McCone's background of engineering and corporate management kicked in when he also announced the creation of the Directorate of Science and Technology (DS&T). In doing so, he forever changed the course of intelligence with the clear mission of advancing the use of science and technology in intelligence collection, analysis, and dissemination.

Wheelon, born in Moline, Illinois in January 1929, had received a BS. Degree in engineering from Stanford University in 1949 and a Ph.D. in theoretical physics from the Massachusetts Institute of Technology in 1952. McCone picked Wheelon for his having spent his professional life in the world of science and advanced technology. Wheelon was at the time working on guidance systems for long-range ballistic missiles and early space projects at TRW, Inc., and teaching at UCLA as a visiting professor.

In his first year, Wheelon met McCone's expectations by fully integrating the Office of Scientific Intelligence [OSI] and the Office of Computer Services into the new directorate. He assumed responsibility for the technical collection and analysis activity of the U-2 overflights and quickly embraced Project OXCART at Area 51 to replace the U-2. At the same time, Wheelon was creating a Foreign Missile and Space Analysis Center and a Special Projects Staff (the future Office of Development and Engineering) to ensure a prominent CIA role at the National Reconnaissance Office, [NRO].

By the time of his departure from CIA in September 1966, Wheelon had created enough of a solid foundation for science and technology within CIA that by 1968, the CIA's Science and Technology Division (DS&T) had turned Area 51 into a high-tech business. The CIA's new Directorate of Science and Technology had, with the U-2 and the A-12, proven the aerial concept. As the agency lost its OXCART intelligence collection to the Air Force, it again took on something that the Air Force could not do. The CIA's new directorate converted its Area 51 to marketing its unique technical services of providing secrecy. Its special projects team at Area 51 began serving customers, mostly the military services, businesses, and agencies seeking to meet the nation's pressing intelligence challenges. The US Air Force and Navy jointly became its first customers.

Like Dr. Wheelon, CIA's special projects team at Area 51 offered many different disciplines ranging from computer programmers and engineers to scientists and analysts. Through the EG&G special projects team of specialists, the DS&T offered other organizations in the intelligence community, the military, academia, the national laboratories, and the private sector its unique tools, capabilities, and expertise at Area 51 to meet and lead in the widely spread of national security challenges.

The CIA's DS&T was a natural to take the lead. After all, it had developed the U-2 reconnaissance plane that dispelled the bomber and missile gaps, and the first stealth plane, the Mach 3 A-12. Elsewhere, it had developed CORONA, the US's first film-return photoreconnaissance satellite, which was crucial to intelligence collection during the Cold War. Now, it was raising a portion of a sunken Soviet submarine from the floor of the Pacific Ocean in a project codenamed, Azorian.

It was not only the Area 51 special projects' expertise that the CIA had to offer its customers. Area 51 was a high-tech aeronautics laboratory where competing aerospace and technology companies knew that could test their product without the competition knowing their business. Area 51 offered its customers secrecy, technical knowledge, services, and timely results on a fixed budget. Area 51 met its customer needs by limiting access to only those essential to getting the job done. Meeting the client's needs would win wars and defend the United States against aggression. If the manufacturer of a new radar avoidance system needed to prove it worked, it did not have to overfly some enemy's denied territory. All it had to do was fly over the CIA's Area 51 to know. Also, the potential buyer of the new system could witness the results in real time. This way, the CIA had two customers, the seller, and the buyer.

The CIA used the U-2 as an example. Eight months after Lockheed's Kelly Johnson signed the contract with the CIA to build the aircraft, the plane took its first flight at Area 51, eight months from scratch to finish and under budget. Furthermore, the U-2 not only completed under budget, but the plane also went operational immediately with the CIA routinely and successfully overflying the USSR to support nation's security needs.

Why Nevada?

The attraction for the CIA to set up shop in Nevada dated back to 31 October 1864. Union sympathizers had expedited Nevada's entry into statehood in the United States to ensure Nevada's participation in the 1864 presidential election in support of President Abraham Lincoln. Because the statehood occurred during the American Civil War, Nevada became known as the Battle Born State.

In 1942, Nevada, the battle born state became the battlefield state when it became a defensive network to repel a feared Japanese invasion of the West Coast. The Navy established its West Coast naval air base outside Fallon, Nevada, and the Army Air Corps established military bases throughout the state. If the Japanese forces invaded the West Coast of the United States, Nevada became the front line. Nevada never ceased its battlefield role for the United States.

The Atomic Energy Commission's (AEC) setting up shop in Nevada to test atomic bombs brought another class of military activity to the state of Nevada. The National Aeronautics and Space Administration (NASA) followed with the NASA High Range, a high-speed flight corridor, and the NERVA Project, a nuclear engine for rocket vehicle application at Jackass Flats. This project developed a nuclear engine intended for a manned flight to Mars. Then came the CIA and Department of Defense (DOD) with Area 51. The entrance of the CIA made Nevada the host to four distinct worlds, i.e., the military world (Navy, coast guard, marine corps, and the Air Force), the white world (Atomic), the space world (NASA), and the black world (Area 51).

A cloak of secrecy-shrouded Nevada's leadership in the national security of the United States. Highly classified activities were occurring throughout the sparsely populated regions of the state in remote areas such as Jackass Flats, Yucca Flats, Yucca Mesa, Frenchman Flats, the Nevada Proving Grounds past the Mercury base camp, and the massive Nellis AFB gunnery range. The Tonopah Test Range tested smart bombs and provided an impact zone for the Regulus cruise missiles launched from submarines at sea.

Apollo astronauts practiced their lunar landings in the atomic bomb craters in Nevada while eight astronauts earned their wings in the X-15 after a B-52 mothership launch in the state. Thus, the CIA could hide its new business in plain sight, as just another classified activity in a venue that thrived off secrecy. Less than five percent of the people affiliated with the projects at Area 51 would ever know of any CIA involvement at this test facility in a barren valley sparsely populated with poisonous, spiny reptiles, insects, silt, and plant life. A desert unmarred by neither hummock nor furrow, no tree or bush growing on the Groom dry pluvial lakebed, its surface of parched clay and alkaline smoothed through the centuries to glass-like flatness from desert winds sweeping water from winter rains across the lakebed in a timeless cycle.

Groom Lake was perfect for what the CIA needed. Its surface was of sufficient hardness for aircraft operations where the entire surface was an active runway under the authority of the Area 51 air traffic control tower. For 13 years, the CIA had operated out of Area 51 in aerial reconnaissance throughout the world. The A-12 alone had flown 1,250 flights out of Area 51 in complete secrecy. Only those with a need-to-know knew of the CIA activities that were occurring in Nevada.

The CIA Did Not Hide the Existence of the Facility at Groom Lake Now Called Area 51.

Other CIA and DOD projects moved in, and the installation's name changed with the projects. Nonetheless, the Atomic Energy Commission, later the Department of Energy, still referred to the Groom Lake facility as Watertown.

For example, seventeen months later, the AEC announced the construction of new facilities for "Project 51." The AEC's M&O support contractor performed the work and provided follow-on support at the site. Because of the requirement for a two-digit area code on the REECo employee timecards, Project 51 morphed into "Area 51" for timecard purposes. Area 51 even had its softball team that played other ball teams in the region. The phone books for the Nevada Atomic Test Site listed Area 51 phone numbers. Local television stations maintained transponders at Mercury, Area 12, and Area 51 to provide television coverage at the remote sites.

1968 following a rain. The water remained on the windward side of lakebed
the lakebed until it evaporated.

1968

EG&G special projects building and radar array for tracking and RCS

RCS Pylon

1968

Adopting the Ways of the Past

During World War II, the Office of Strategic Services (OSS)—the predecessor of today's CIA—was the first to use science and technology as part of the intelligence process. The OSS's Research and Development (R&D) Branch invented weapons and gadgets and adapted Allied equipment for new missions. Mostly, however, the CIA dealt with covert activities using silenced pistols, tiny cameras and such things as a uniform button containing a compass. These were a few of the items created by the OSS R&D Branch.

In 1947, when President Harry S. Truman founded the CIA, many R&D veterans went to work at the agency, bringing with them the latest scientific advances. They also brought with them the black world use of pseudonyms, aliases, and code names. Organizations hid behind false names and vague terminology. Thus, for the past 13 years, the CIA had kept its science and technology operations at Area 51 secret using pseudo names and call signs. The successes of the U-2 program managed by Dick Bissell and inspired by technical people rather than covert operatives had convinced the upper management that it needed a new Directorate that thrived on science and technology. The agency suddenly controlled what was going on around it. With this control came responsibility unlike that of covert operations in some war zone or banana belt country. Instead of running operatives, the Area 51 station was running operations with tentacles extending around the world.

One of the first things the agency did with Project OXCART was set up a Special Operations Area (SOA) to shield refueling area that the A-12 required immediately after takeoff. It identified this controlled airspace around Area 51 with the name YULETIDE. Now, with the A-12s gone, the airspace around Area 51 needed a secure air space for another reason. That was the Soviet MiG planes conducting local flights while being exploited at Area 51.

The control tower used call signs BUD and JUPITER. The pilots knew approach control as SAUCY, and DUTCH CONTROL DUTCH CONTROL or BOXER CONTROL handled A-12 flight operations depending on the radio frequency used.

The A-12 pilots (called "drivers") used DUTCH call signs, as well. In the early days of the program, everyone associated the call sign with the Article number, so that a pilot flying. Article 123 identified as DUTCH 23. The pilots, later, used their assigned personal call signs regardless of aircraft number. Frank Murray, for example, used DUTCH 20 and Robert Gilliland, the Lockheed test pilot who first flew the SR-71 used DUTCH 51. The CIA's first customers, the pilots now arriving with the MiG planes adopted the name, "Bandit."

Adjusting to More Than One Tenant

Previously, until January 1968, the CIA Groom Lake station had only one operational project, and everyone there supported one tenant — the Lockheed A-12 Project OXCART to develop a replacement for the CIA's U-2 spy plane. Area 51's single project status changed in January 1968 as Project OXCART wound down and with the arrival of Project HAVE DOUGHNUT, a joint USAF/Navy technical and tactical evaluation of the MiG-21F-13.

The purpose and needs of this initial MiG exploitation project at Area 51 drew an entourage of interested and participating intelligence and specialized agencies, all following the lead of the US Air Force and Navy. The services provided to these new customers by the CIA far exceeded the preceding RCS and flight tests during Project OXCART, then the only tenant.

Suddenly having multiple customers also changed the way the CIA conducted business at the station. The customers arriving to work on the new MiG project did not have a need-to-know about the A-12 project and vice versa. Access to the different sections of the station suddenly became restricted, as did social communication and networking at the agency's flight test facility.

The need-to-know and compartmentalization also applied to the members of the special projects exploitation team functioning as cadre for Area 51. For the first time, some on the team worked for one customer and not another.

Housing suddenly segregated with each compartment establishing their own social and entertainment centers and activities. The residents now muted any conversation related to work activities when they entered the shared mess facilities. The participants of the two projects suddenly found themselves mixing with personnel lacking the need-to-know of personal identities and the project they represented.

When Project OXCART ended in June 1968, so did the services of the 1129th Special Activities Squadron whose only existence was to support the CIA's A-12 program.

The customers now showing up for the MiG exploitation project hosted by the CIA at Area 51 included Foreign Technology Division FTD) leading an exploitation team of specialists drawn from throughout the USAF and USN. For the Air Force, the participants included AFSC (Air Force Systems Command), the Laboratories at Wright-Patterson AFB, the AFFTC (Air Force Flight Test Center), SAC (Strategic Air Command), and the TAC (Tactical Air Command), and the Aeronautical Systems Division (ASD) responsible for equipment procurement and sustainment. The NATC (Naval Air Test Center), and the NWC (Naval Weapons Center) participated for the Navy. The NDA (National Defense Agency), and the NASIC (National Air and Space Intelligence Center) represented the various intelligence agencies.

The CIA commanders at Area started with Richard A. "Dick" Newton USMCR (Ret) (1955–1956) for the U-2 program until replaced by Landon McConnell (1956–1957). Werner Weiss had arrived in 1960 for the A-12 Project OXCART and would not leave until 1969. Richard A. "Dick" Sampson, formerly embedded with the U-2 program at the Skunks Works, arrived at Area 51 to replace Weiss in 1961. He remained until 1971 when Sam Mitchell became the CIA station commander for the unconventional operations occurring at and out of Area 51.

Area 51 Special Projects Exploitation Team Cadre

The termination of CIA Project OXCART in early 1968 drastically reduced the number of special projects personnel. The RIF took away key personnel such as Harry Phiffer, the one in charge of the engineers at the site. Wayne Pendleton Flight-Systems (flight) manager went to work for Lockheed, and Jim Tarver G-Systems (ground) manager took an assignment elsewhere. Jules Kabat, responsibility for the two radar systems in the antenna building, the S-band radar with the 60-foot dish and a Navy radar left the Ranch to accept an assignment in Vietnam.

Those chosen to remain for the future MiG exploitation and stealth projects included key personnel such as:
- Thornton D. Barnes- (Nike radar),
- Dick McEwen-(G-systems),
- Cowan Dawson- (C-band radar),
- Dave Haen-(Q-bay),
- Sam Gamble-(Draftsman),
- Jim Freedman-(Admin),

The Carco C-47 pilots out Of Albuquerque, New Mexico, were: Roy Kemp-Chief Pilot, Tom Hall, Joe Cotton, Tom Losh, and Flo Deluna-aircraft mechanic.

The CIA preferred keeping men married with two children and common interests. The agency felt this created the necessary bonding for sharing national security concerns. It did. It developed a cohesiveness where they worked together all week and then played together at the lake or on the mountain during the weekends. Rarely did they talk shop, and even under these conditions, they never did so with the spouses or children present.

Each member of the special projects team and their families had undergone the same security and compatibility evaluations, though many of them evolved from the atomic testing side. Consequently, almost every member the Special Project team carried both an AEC Q security clearance and a DOD top-secret security clearance.

Babjack, R. J.	Freedman, Ralph J	Kirchhoff, Robert T.	Starry, Clifford E.
Barnes, Thornton D.	Gamble, Sammie L.	Leonardi, John	Thomas, Jeff D.
Beahm, Glen M.	Grace, John W.	Long, James E.	Vittetoe, Dennis E.
Becherer, Charles B.	Haen, David B.	Luker, Bobby V. M.	Washam, Charley P.
Christensen, Calvin D.	Hardy, Leroy C.	McGlothen, Willie	Watson, Galen E.
Dawson, Cowan F.	Hunt, Lee D.	McLeod, William F	Weed, James P.
Dockter, Marvin R.	Heaps, Kenneth L.	Owens, Elridge W.	
Evans, Paul M.	Jenkins, Wesley G.	Swenson, Marvin L.	

TD Barnes   Ralph (Jim) Freedman   Dave Haen   Denise Haen

With Project OXCART, the military sent in officers and enlisted men specializing in such support needs as medical, construction of fuel systems, and supply. The military pilots did not arrive until near four years into the program when the A-12 was ready to fly. The Air Force arrived to maintain and fly the F-101 VooDoo chase planes. They maintained the facility with duties outside those of the specialized activities of the CIA, EG&G special projects, Reynolds Electric, and Lockheed Aircraft Co. The Air Force instructor pilots, who learned to fly the A-12 trainer from the Lockheed test pilots, in turn, trained the CIA pilots to fly the A-12. No Air Force pilot, however, ever flew an operational A-12 plane, nor did a CIA A-12 pilot every fly the Air Force's SR-71 while still employed by the CIA.

Now, as the A-12 planes retired into storage, Area 51 no longer needed many of the services remaining from the past. The CIA and Area 51 no longer needed the mission planners, or the technical support personnel such as Pratt & Whitney, David Clark, Co. Sylvania, Hycon, or Perkins Elmer, to name a few who maintained the A-12 aircraft and pilots requiring pressure suits.

Following the early 1968 closure of its Project OXCART, the CIA reduced the size of the special projects to 31 select members known administratively as Organization #6300. Most of those not retained returned to EG&G's atomic activities at the Nevada Proving Grounds now known as the Nevada Test Site. No longer could the prime contractor, EG&G fill an open slot with just anyone from the company's atomic side. The CIA ensured that EG&G established a support cadre based on the individual's unique qualifications, family stability, ethics, integrity, and moral qualities as well.

No women worked on site. (Denise Haen worked out of the Las Vegas office.) To that extent, initially, the Air Force even denied concurrent dependent travel for the Air Force support personnel assigned to Area 51, classifying it as an unaccompanied tour. Those in the 1955-56 U-2 program found the no dependent policy acceptable because all concerned knew they would leave once the training stopped. Area 51 was a temporary duty in those days. Now it was not.

The transient aspect all changed in 1968 as the CIA's science and technology division geared up to focus about their name, science and technology. Its purpose was no longer transitional as to projects. Gone were the operations mindset, replaced by service – customer thinking.

The United States was going through a transitional time. Congress had just authorized the US Metric Study, a three-year study of systems of measurement in the United States, with emphasis on the feasibility of metrication. Parents were screaming about not understanding the New Math that was now part of their children's school curriculum. The CIA had mastered the Mach 3 speed idiosyncrasies and was now looking at the challenges of stealth. The CIA and special projects staff were no longer there for the duration of a project. The special projects team was building a computer specifically for its Area 51 science and technology activities. Area 51 was there to stay. The S&T was articulating from slide rule technology to that of computers, vacuum tubes to transistors.

In December 1963, the support of the EG&G special projects at Area 51 had become more indispensable when it gained access to the AEC's 1604 computer. With a half rack of equipment, the special projects could check the CIA's radar cross-section data in hours, saving the agency having to transport it back to headquarters for reduction and analysis. Consequently, the CIA had agreed to the installation of the extra equipment and authorized the EG&G to use the computer facilities for the more routine work. The CIA would still do the data reduction and analysis in those cases where more advanced areas of work needed EG&G's services, which at the time was putting equipment and programs together to simulate multiple targets against a SPOON REST radar. Now, with Area 51 changing to focus on science and technology, the special projects team did in fact design and built a special computer for the upcoming needs of Area 51.

When Area 51 had only one tenant and purpose—that being Project OXCART, everyone had the run of the facility, most still unaware that it belonged to a CIA station. Many of those serving and unique to Project OXCART left Area 51 at the same time as their project, leaving behind the specialists needed to advance Area 51 technology into the future. Now, with multiple customers, the special projects and CIA personnel housed and transported separately from any other tenants. People could not merely walk into some building out of curiosity to see what was going on. Suddenly, members of the special projects team were reporting only to their customer and no longer talking shop with each other.

The addition of new highly classified projects tightened the compartmentalization and need-to-know environment within the special projects exploitation team. With Project OXCART, the special projects team worked on individual projects and for entirely different purposes. The arrival of the MiGs split the focus of the special projects team to those specializing in advancing stealth technology and those technically supporting the exploitation of Soviet technology.

By the special projects team members not talking about their work with each other did not mean they did not want to get to know each other. Working in such secrecy bonded the families. None felt comfortable socializing outside their team. They did not trust strangers. The result for the men and their families was their isolating from the outside world and bonding as a family.

Most understood in basic terms what the other did from giving them a hand with something. Nonetheless, they did not ask what, why, who, or how if they lacked the need to know.

The isolating did not mean the team members lost their curiosity. Quite the contrary, however, each member, with his specialized ambitions and professional goals, expected protection from snooping regarding his projects. The members assisted each other as needed. They, however, totally respected the proprietary rights of each other's thoughts and achievements.

These highly motivated workers did not work by the time-clock. The CIA selected each based on his pristine ethics, his needed specialty, and his drive to achieve the impossible. Most pilots at Area 51 considered flying an opportunity and not a job. It was the same with the special projects team. Those selected did not consider what they did at Area 51 as work — they felt it was a duty, privilege, and pleasure. It too was an opportunity to advance one's ambitions, his vision for advancing science in his field.

The Central Intelligence Agency operated the Area 51 facility, yet, few of the workers or customers knew this. They had no need to know. The special projects team knew of the CIA affiliation because the team members reported directly to the agency. This did not mean that they knew the identity of the customer. Again, they had no need to know. Even within the special projects team, the team members might know each other's skills, but not their background — what brought their contemporaries to Area

51. Each of them was there because of having a background, and skills unique to the needs of the CIA.

As an example, Dave Haen specialized in telemetry during Project OXCART. During Project HAVE DRILL, he became involved with a new radar system. As with most of the special projects team, from that point on, his career remains classified top secret. Over the course of over 30 years in special projects, Dave advanced to Director of Site Support Operations. Dave married Denise, EG&G's contracting officer previously mentioned as having inspected the facility at Area 51.

Jim Freedman came from the atmospheric testing of the atomic bomb in the various remote areas of the world to the special projects team as an administrator and courier. Each evening, Freedman dropped by CIA commander's office to pick up the dispatch to Langley and dropped it off to a United Airlines employee at McCarran International Airport in Las Vegas. The next morning, Freedman repeated the procedure in reverse, delivering the dispatch from Langley to Area 51. While the rest of the special projects team stayed the week at Area 51, Freedman made his daily courier runs to Las Vegas.

Sixty-five miles west of the Area 51 facility, TD Barnes had conducted the Seven Sisters tracking of the A-12 and YF-12 for the CIA from the NASA tracking station at Beatty. Operating the NASA High Range did not require a high-security clearance; however, Barnes having recently left the military still held the required security clearance. The agency knew Barnes from his previous involvement with the CIA related to Project OXCART during Project PALLADIUM.

He, like most of the others, collected for the team, had the specialized training and experience needed to operate Area 51 as the CIA intended. He had graduated after attending over two years of schooling at the Nike Ajax and Nike Hercules surface-to-air radar and missile maintenance schools at Fort Bliss, Texas. He'd also graduated from the six-month HAWK surface-to-air missile maintenance school at Fort Bliss, doubling up on the specialized radar experience needed at Area 51 for airborne high-speed RCS radar cross-section evaluations. He brought with him several years' experience at tracking 3,000 mph missiles and was one of only half a dozen in the world experienced in tracking the Hypersonic Mach 6.7 X-15 rocket plane using NASA's SCR-584 Mod-2 radar at Beatty, Nevada. Giving him, even more, diversity, his job at the NASA radar site included other systems besides the radar, systems such as telemetry, DTS (data acquisition and transmission), analog to digital conversion, timing, and communications. Barnes was unique in that he was familiar with NASA's SCR-584 (short for Signal Corps Radio # 584) microwave radar developed by the MIT Radiation Laboratory during World War II. This was the same radar system used by the Soviet Union against the agency's U-2 and A-12 reconnaissance planes. He had gained even more experience with it as a member of the Seven Sisters, a classified element of CIA Project OXCART and the CIA/Air Force Project KEDLOCK that included six Air Force SAGE ADC (Air Defense Command) radar systems.

Working at Area 51.

Almost all engineers in the special projects team chose a senior technician classification, allowing them to escape the "salaried" pay status of an engineer. As senior technicians, the special projects team drew almost the same pay as engineers. However, after the first eight hours, their pay rate increased to time and a half for four hours. Their pay scale then went to double time straight through, 24 hours a day until they arrived back in Las Vegas on Friday evening.

They received free food and lodging and even provided with snake proof boots for when they worked at the pylon out on the lakebed.

Area 51 deputy commander Werner Weiss and his successor, Dick Sampson of the CIA, accommodated the permanent party personnel at Area 51 with a small BX. It stocked snacks and various personal hygiene items. Other amenities included a swimming pool, exercise room, softball diamond, putting green, pool room, and a three-stool bar called "Sam's Place," named after Dick Sampson.

The amenities at Area 51 reduced the hardship on the special projects team members maintaining two residents. Area 51 provided each room in a row of Babbitt housing units. Each house contained a small living room where the team played poker and watched 8 mm movies played on a film projector.

The special projects group usually banded in two groups for housing, one group of the boating enthusiasts with boats moored on Lake Mead and the other as the Mt. Charleston cabin dwellers. The CIA emphasized personnel selection including common interests, not only with the workers, however, their families as well.

When Area 51 was reactivated in late 1959 to begin flight testing of the A-12, the site was administered by the 1129th Special Activities Squadron with Col Robert Holbury in command. His deputy and the agency's chief of station was a senior agency representative, Werner Weiss, a GS-15 with the Central Intelligence Agency. The CIA facility was composed of US Air Force personnel, civilian staff, and contract personnel related to the A-12 project, and the EG&G special projects contractor personnel engaged in CIA project research, the latter reporting to the agency's science and technology division at Langley.

Reynolds Electrical and Engineering Company (REECo), under the direction of the CIA management, performed site maintenance and housekeeping and supplied the facility with everyday housekeeping and maintenance supplies and equipment under the general administration of the Area 51 Base Commander.

The CIA Station, Area 51, was located approximately 120 miles northwest of Las Vegas and 53 miles north of Mercury, Nevada, at the edge of the Nevada Test Site. It could be reached only by automobile and aircraft., Constellation from Los Angeles and C-47 from Las Vegas. Since there was no public surface transportation between Area 51 and Las Vegas, and since space on the contractor C-47 was available only on a first-come, first-served basis, personnel were urged to bring their automobiles since it may be their only means of transportation between the Area and Las Vegas.

The cost of living in Las Vegas was higher than the national average, probably one of the highest consumer indexes in the US. During their stays at the station or flight test facility, the CIA provided all personnel at Area 51 assigned quarters and provided meals without cost regardless of them having a family in Las Vegas or another city. Personnel lived in Babbitt houses or trailers, one or two men to a room. All homes and trailers were well furnished and had air-coolers and heaters. The agency provided towels and bedding, including linens and blankets, without charge. It advised incoming personnel that it provided only simple field living conditions with no locked storage suitable for the protection of valuables within the Area. Therefore, personnel had to provide their footlockers.

The agency also requested its people not to bring TV sets as it provided television sets in all houses and trailers and personnel. Furthermore, the antenna arrangements did not provide for unlimited hook-ups. Area 51 received poor AM radio reception during daylight hours and no FM reception at all. Those considering a transistor-type radio needed no less than a 7-transistor unit.

Reynolds Electrical and Engineering Company provided excellent messing facilities. The newly constructed mess hall managed by Murphy Green accommodated approximately 1,500 people, and hours of operation assured all personnel, regardless of working shifts, three full meals daily. There were no shortages, and fresh fruit, milk, coffee, tea, pies and cakes, salads and ice cream were always on the serving line.

The weather was ordinarily extremely dry, although the valley experienced severe rainstorms and occasional snowstorms. The high desert area anticipated extreme seasonal temperatures with below-freezing temperatures during the short winter months and extreme heat and intense sun during the summer months.

The facility had a free do-it-yourself laundromat, with soap and bleach furnished. REECo delivered laundry to a Las Vegas laundry on Tuesdays and Thursdays weekly and picked it up a week later. Dry cleaning laundry was expensive, for example, it cost about 85 cents to have a pair of trousers dry cleaned and about 45 cents to have a shirt laundered.

Informal wear was the order of the day during working hours. The standard attire during the summer months included walking shorts and T-shirts or sports shirts. With few exceptions, all buildings were air-cooled or air-conditioned. The temperature reached 120 degrees outside.

For medical facilities, a United States Air Force staffed dispensary provided 24-hour medical coverage for the entire facility. The facility had no hospital facilities, so military personnel was authorized to use Nellis AFB facilities in Las Vegas while civilian personnel had to use the hospital in the Las Vegas or Los Angeles area. Nellis AFB provided a clinic for military personnel and their dependents assigned to the detachment every Friday by the Air Force officer general practitioner assigned to this station.

For recreation, the station's a newly constructed conditioning building contained six bowling alleys installed by Brunswick Corporation, a dry heat room, enclosed swimming pool, handball court, squash court, physical improvement room, paperback library, and a basketball court. The bowling alleys came equipped with automatic pinsetters and ball returns and charged a fee of 25 cents that became slightly higher priced for higher-grade movies.

The recreation hall contained a beer bar, hardback library, six pool tables, a TV room, hobby room, shuffleboard, and a snooker table.

The full-size gymnasium contained a basketball court, badminton court, and an exercise area. Outdoors was a softball field, four tennis courts, a basketball court, a volleyball court, a badminton court, a hand court, and a gold cage. CIA pilot Frank Murray and others took up flying model planes, and many of those at the station later became Ham radio operators.

All privately owned autos required registering with the Area 51 security office. Auto insurance rates were extremely high in Nevada.

For this staying at the station over weekends, Camp Mercury and Indian Springs AFB held religious services. Transportation was made available to anyone desiring to attend the service of their choice.

For barbershop services, facilities were available at Camp Mercury, Las Vegas, and Nellis AFB. Every Wednesday, there was a barber at Area 51 from 0800 to 1800 hours. The average price of a haircut throughout southern Nevada was $2.50. However, military personnel could get haircuts at Nellis or Indian Springs AFB for $1.25.

The station offered a small on-site PX in the recreation hall, offering toilet articles, tobacco, magazines and newspapers, soft drinks, and beer.

Area 51 offered no public telephones on site for personal calls, however, the Area 51 military commander could authorize emergency calls from on-site.

For mail service, those at the station used the first-class post office at Camp Mercury. The civilians residing off-site used P.O. Box 732, Las Vegas, and military personnel used P.O. Box 882, Las Vegas. Both provided daily service and mail delivery to the Area.

The station allowed no animals or pets. It allowed personal cameras, but prohibited photography within any section of the Nevada Test Site, including Area 51. Upon arrival, all cameras required registering with the Area 51 security office. The Nevada Test Site prohibited firearms of any type. The CIA allowed transporting firearms to the Area provided the carrier immediately registered them with the Area 51 security office.

Lizards, copperhead and king snakes, deer, mountain sheep, jackrabbits, bobcats, and badgers inhabited the surrounding desert area in ample numbers.

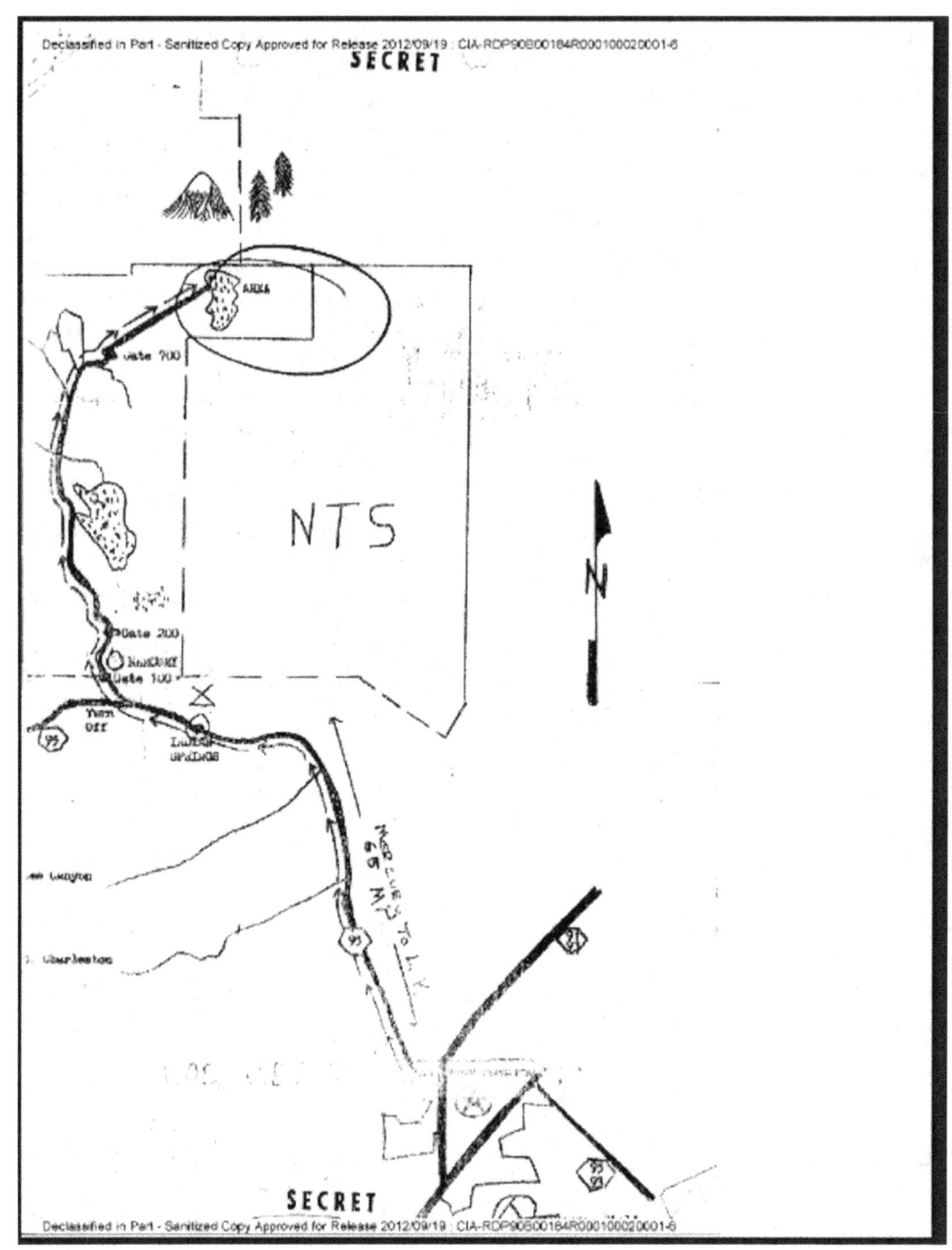

Automobile Route Through the Nevada Test Site to Area 51

Groom Lake Mess Hall circa 1960 - replaced with one seating

Babbitt Housing brought in from the Naval Warfare Center at Hawthorn, NV

Approved For Release 2002/06/24 : CIA-RDP75B00326R000200120005-6

OXC-4560-63

27 FEB 1963

MEMORANDUM FOR THE RECORD

25X1A

SUBJECT:

1. On 12 October 1962, the undersigned, in the company of Messrs Cunningham, [   ] and General Flickinger participated in the recruitment of subject for Project OXCART.

25X1A

2. The first day followed pattern previously established and went pretty much according to script. Individual was initially interested in performance characteristics and, although his detailed questions were parried, he did obtain an admission that speed, on the conservative side, might be said to exceed MACH-2.

3. During the interview, one could see that General Flickinger was not impressed with Subject. This stems from concern over very inquisitive, intelligent, and domineering wife. Secondly, physiological findings, although disqualifying, which pertained to spatial bifida were a cause for concern as earlier raised by Colonel Ledford and General Flickinger.

4. On 16 October 1962, [   ] and undersigned met Subject at the Shoreham Hotel for an answer. He indicated a great deal of concern about hiding the truth from his wife and, although it was first thought he was using this as an excuse, as the interview progressed, it was quite clear that this would be a source of emotional stress. This, together with the other considerations, placed [   ] and the undersigned in the position of disuading the individual from volunteering. At the conclusion, he was actually upset, but believed that his declination was made by himself and was in the interests of all parties concerned.

25X1A

25X1A

5. Subject was given a security admonishment by [   ] of not getting trapped into divulging what had transpired. Subject should forget that this recruitment effort had taken place. Termination security oath was executed by Subject. Subject could foresee no questions which he would not be able to handle. He will indicate that he had withdrawn from the voluntary program because it was not leading anyplace. Subject further indicated that there would be no need to explain why he had brought his flight gear on this trip.

25X1A

25X1A

Chief, Personnel Branch

Approved For Release 2002/06/24 : CIA-RDP75B00326R000200120005-6

SECRET

The CIA looked at the entire family when selecting those to work at Area 51. A spouse with a drinking problem, sleeping around, or gambling excessively could jeopardize the worker's security clearance. Worse yet was a wife who could not handle her husband being gone all week and not telling her where he was or what he did. As the document above shows, if an applicate had to get approval from his wife to work at Area 51, he was not considered for the job.

Further grouping often occurred to segregate the team, per project or customer. Extracurricular activities such as poker, rental movies, reading, and so forth, developed, even more, sub-grouping per interests. (Note about these recreation activities: Near Las Vegas, Lake Mead, the largest reservoir in the United States lies along the Colorado River. Mount Charleston, officially named Charleston Peak, altitude 11,916 feet, lies about 35 miles northwest of Las Vegas),

CIA and other personnel staying at Area 51, such as Air Force, Lockheed, Hughes, Pratt and Whitney housed in similar houses, however, clustered apart from the others. Hardly any association existed even in the special projects team outside one's professional group.

For their off-duty pleasures, each customer developed their entertainment, which involved converting a room in one of the residential houses into a bar and poker room. The OXCART group had what they called, "House Six." Those interested in golfing prepared a three-hole golf course. Others took up model plane flying, and many Ham radio operators emerged from their tour at Area 51.

An exception to the commingling of groups came with the forming of the Area 51 softball team that competed with a softball team formed on the Atomic Testing Grounds.

The permanent party personnel mainly used Sam's Place and the CIA amenities, except for poker games conducted in individual houses, which typically offered four bedrooms, a shared living room, and kitchen.

The approved driving distance from Area-51 through the Nevada Test Site to "town" (Las Vegas) was 120 miles. The REECo personnel, mostly laborers, cooks, housekeeping, carpenters, and electricians drove in and out.

Base security allowed one Lockheed radar cross-section tester to fly his Cessna to and from the Antelope Valley. In exchange, he assisted in security sweeps, even installing a failed U-2 window installed in the bottom of the plane to enhance ground viewing.

The CIA required all customers, most Air Force personnel, Air Force civilians, contractor maintenance personnel, tradecraft people to commute in and out on C-47s and Lockheed Constellations.

The CIA personnel drove to the site. EG&G special projects personnel flew to Area 51 on CarCo, a subcontract airline of the Atomic Energy Commission.

The airplanes for EG&G started small and got larger as the project also grew. Initially, they flew to and from the site in a D-18, then a C-47, and then an F-27.

Each group bore their incidences of excitable moments. A C-54 had crashed on Mt Charleston early in the U-2 program killing 14, all aboard. Those killed included an Air Force crew, key CIA, and Hycon employees.

Grouped even further the author and three others on the special projects team flew to and from Area 51 in a Beechcraft Queen Air flying out of a secure compound along the side of the McCarran Airport along Sunset Road. Though unconfirmed, some believe that Central Intelligence mainly chose the Queen Air group for their unique contributions, looking ahead to future projects rather than merely the current project.

The mess hall, open 24/7 remains to this day the major and longest living amenity at Area 51. Anyone ever serving at Area 51 lays claim to Area 51 having better food than any hotel or casino in Las Vegas. Having a world-class mess hall offset the severity of the regimen.

Tuesday and Thursday night dinners, famous for the steak baron of beef and lobster drew authorized support personnel from Burbank and all over who scheduled their trips to the area around those days.

The Air Force support element during OXCART arranged with the commissary officer at Wendover Air Base, Nevada to provide the mess hall with fresh oysters and lobster. Evidence of the great days of the CIA era remains today in the huge piles of oyster shells dumped on the Groom dry lake from the mess hall. Future archeologists will wonder how and when Area 51 connected to an ocean.

Circa 1968, installation of a translator finally brought television to Area 51. Instead of providing a form of relaxing entertainment, it invoked outbursts of disgust. The Area51 residents might be watching a great game of football, only to have the stronger signal of a bullfight out of Mexico override Channel 3 out of Las Vegas. The TV signal propagation changed with atmospheric conditions, causing skip conditions. Meanwhile, the poker games continued as the prime entertainment.

———

During Project OXCART, the Groom Lake evolved from a high desert surrounding a large dry lakebed surrounded by semi-barren mountain ranges within the Emigrant Valley to a fully functioning airbase. Consequently, as Project DOUGHNUT moved in, it did not require installation or construction, leaving little for the permanent party personnel to do other than their specialty.

Project OXCART caught the spying eyes of the Soviet Union, bringing satellite overflights of what Area 51 personnel dubbed as ashcans. However, the arrival of the first MiG drastically increased the Soviet interest in the activities of Area 51, resulting in an increased frequency of satellites passing overhead. As with OXCART, the passing satellites required the ceasing of any electronic emissions and the hiding of any outdoor activities. That included rushing any planes caught on the airstrip or tarmac into a hangar or hoot-n-scoot shed existing for that purpose.

Besides severely disrupting schedules and activities, the downtime because of passing satellites created a toll on the CIA's highly skilled and highly motivated personnel on the special projects team. To most, the inability to energize any signal emitting electronic equipment made their assigned project unbearably dull. This boredom, however, unintentionally developed new technology from the special projects personnel using their knowledge, curiosity, and available means to experiment with something that later became a project at the facility.

Sadly, these achievements and most of the individual human accomplishments, mistakes (referred to as OS or "old shit!" moments), and humor have and may never become known.

During OXCART, the Lockheed engineers raised the nose of the first A-12 Blackbird, using a forklift, to evaluate fuel distribution. The fuel ran to the aft end of the plane, lifting the nose of the 105-foot plane into the rafters of the hangar. OS!

A long-running incident attributable to the special projects team is one referred to merely as "the goat incident."

Following the rain, the Groom dry lake bed always retained water for long periods. The lakebed did not absorb the water, so it stayed until it evaporated.

The goat, an all-terrain, six-wheel drive, an amphibious vehicle so named by the special projects team, hauled electronic test equipment and tools to the radar cross-section pylon on the Groom lakebed. During a trip to the pylon, the vehicle somehow caught fire, becoming well advanced before the operator noticed it burning in the rear of the vehicle.

The fire destroyed the goat and its contents. For years after that, any time a piece of test equipment became lost or misplaced; the team immediately attributed it to the loss of the goat. Later, an inventory of all the equipment supposedly lost with the goat revealed losses to fill a small truck. (Most of the items declared as lost usually showed up in use by someone else at the time).

Background:

The special projects personnel, activities, and equipment discussed herein occurred half a century ago and did not suggest the same existing today. The author emphasizes that this book discusses what applies to the declassified projects only.

The CIA knew the Russian radars tracked Gary Powers et al. flying missions over Russia in the U-2. Altitude saved the U-2 from planes and missiles. The CIA also knew the U-2's vulnerable, which Gary proved.

The CIA needed a replacement and upgraded aerial vehicle if it intended to continue overflying Russia. The CIA wanted the replacement to have a low radar cross-section. Kelly Johnson added speed and altitude to the replacement plane, the A-12.

All aircraft bought by the government had requirements and specifications which require verification. A company called SEI (Scientific Engineering Inc.) claimed a method of doing that for radar cross-section using a radar ground range and a radar flight range.

The Groom Lake flight test facility rated as the best place to conduct radar cross-section evaluations. However, SEI needed help to construct such a big project. To this end, EG&G employed a pool of Q cleared (AEC Top Secret) personnel, which helped the security aspects. EG&G provided some engineers and technicians from its nuclear activities next door at the Atomic Energy Commission atomic bomb testing.

Having this equipment in place and operational at Area 51, plus the security it offered made it the obvious venue for the MiG exploitation projects.

Some of the equipment used by the special projects team to control, receive, process, and store data from the tactical missions.

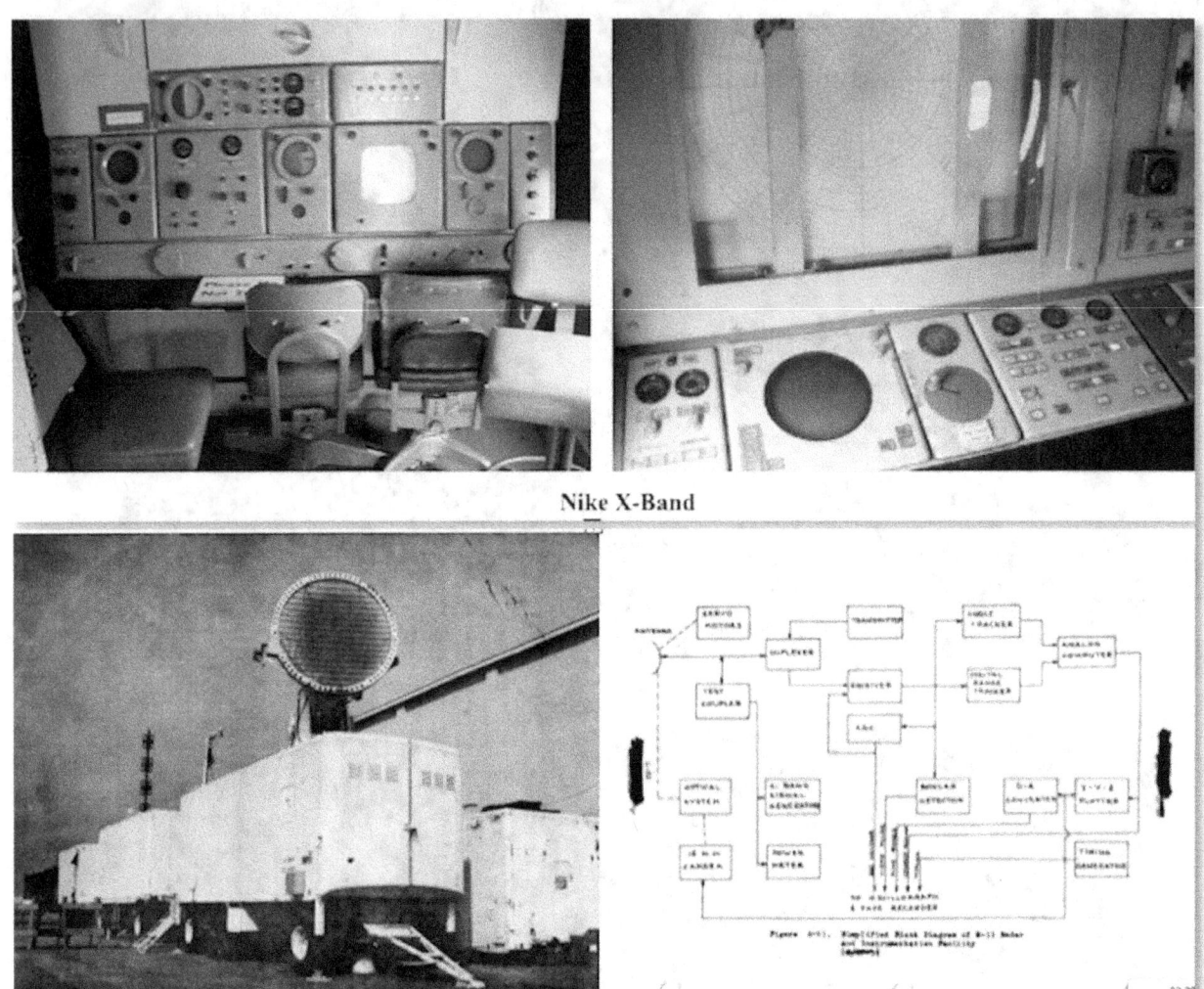

Nike X-Band

Control Console #1

Power Supplies → Nike Bullgear

Control Console #2 X-Y Plotter → Data Recorders & Storage

The Special Projects RCS site.

For airborne RCS tests of the A-12, the Lockheed Flight Test Director called the EG&G special projects team for mission briefs. These training flights typically flew to Mountain Home, Idaho and back, passing overhead before landing.

At first, the special projects team found it difficult to locate and lock onto the plane traveling inbound

31

at Mach 3. The radar operators searched for the return of the aircraft by scanning the horizon with the Nike radar used to track it on the inbound leg. Once the target appeared, the operator had only seconds to lock the radar onto the target. Heading inbound at Mach 3, the range decreased, and the elevation increased at too high a rate for the radar to slew and catch up.

The Nike radar operator acquired target acquisition by sweeping his beam at the expected azimuth for the target to pop over the horizon. With the radar azimuth and elevation locked on the target, the target's transponder answered with a spike for the operator to capture and lock in the range gate.

Once the Nike radar acquired a target lock, it drove the bull gear to align all other antennas on the target.

John Grace solved the target acquisition problem by bringing Barnes in from the NASA High Range at the Beatty Radar Station. Barnes, a hypersonic flight support specialist routinely locked onto the much faster X-15 and tracked it as it passed directly over his radar site at Beatty. (On the NASA, High Range in Nevada, Barnes maintained his proficiency with the radar system, the data transmission, communications, microwave, and telemetry systems. He participated in tracking the Mach 3 A-12, YF-12, and M-21 Blackbirds, the Mach 3 XB-70, and the Mach 6.7 X-15.)

A typical airborne radar cross-section of the Mach 3 A-12 started with the A-12 headed away from Area 51. The Nike radar tracked the aircraft with the other radar systems involved passively ganged to the Nike servos. The objective was to obtain a basic radar cross-section under operational conditions. However, they never tested the electronic countermeasure (ECM) or electronic warfare systems equipment. The aircraft had a special paint treatment.

The flight plans called for the radar slant range versus elevation angle for the approach portion of the flight. For the refractive index model selected, the altitude of the aircraft appeared to be between 72,000 and 75,000 feet, about the radar test site, for the inbound run.

The Nike radar locked the Article in auto track outbound at nine nautical miles and tracked the plane to 255 miles where it became inbound. They plotted the ground range versus the azimuth range as the aircraft approached the radar station from 340 degrees.

The central reference servo system (CRS) plotted the rapid increase in elevation angle as the plane approached. The Bendix system was operated in the monostatic configuration to obtain cross-section data for horizontal polarization in S-band (21.73-2.98 GHz). A portion of the cross-section data was invalid because the maximum slew rate of the DSK was insufficient to provide antenna tracking as the vehicle passed nearly overhead.

The Navy system was operated in the monostatic configuration to also obtain cross-section data for horizontal polarization at 173 MHz at a range of 140 nautical miles. The plot of cross-section data versus time for the Navy system indicated that the cross-section varied widely, with a maximum of 14 dBam. Again, the DSK was insufficient to permit antenna tracking as the vehicle passed overhead.

The MOD III system operated in the monostatic configuration to obtain cross-section data for vertical polarization in the c-band (5.47GHz). Like the other cross-section radars, the MOD lost track as the A-12 flew overhead. Nonetheless, it obtained values as high as 18 Db above 1 square meter for the cross-section of article 132 at C-band.

Once the A-12s deployed to Kadena for Operation BLACK SHIELD, the range shut down in June 1967 and remained so until early in 1968 when the MIG-21 arrived for exploitation. It was this experience learned while testing the U-2 and A-12 that qualified the special projects team for their role in exploiting the Soviet MiGs.

The CIA brought considerable knowledge and expertise to the table when the Soviet MiG planes arrived at Area 51 for exploitation. The methods and equipment used for the previous U-2, and A-12 projects and the technology gained applied to the MiG exploitation projects as well.

The RCS goals defined in the primary contract for the A-12 required two radar ranges, ground, and a flight range. The RCS team measured a U-2 on the ground range to verify the correlation between ranges and then flew and measured it on the flight range for comparison.

The ground range assisted in developing the A-12 configuration and defining its radar cross-section (RCS). The flight range verified the RCS of the A-12 in flight.

The ground range extended 1-mile long north on the southwest edge of Groom Lake. A major positive for using the lakebed, it provided a low noise background for the radars.

The 1-mile site supported an actual A-12 as well as models for measuring the RCS. A ½ mile site using an inflatable airbag to hold scale models and model sections of the A-12 to measure the RCS. A ¼ mile site consisting of a square meter sphere which erected from below ground to establish calibrations of the radars to a 1 square meter target. (0 Dbsm)

The radar antennas on the lake edge faced down range to the North. The heights of the support poles/airbag/cal sphere correlated with the radar antenna heights to maximize the radar signals. Special projects mounted the radar targets upside down and rotated them while the radars illuminated them. They recorded the return signals and referenced them to the calibration sphere.

The one-mile site contained a hydraulic pole buried in the lakebed. The pole, a couple of destroyer propeller shafts welded together, raised the targets up 52 feet. A 26-foot cube of concrete below the lakebed supported the pole. A hinged parabolic cover bolted to the top of the pole reduced the radar backscatter of the pole. The ground crew at Area 51 mounted the target with an internal rotator, mounted on the pole at ground level and then raised it. The ground crew then wrapped the cover around the pole to minimize the pole RCS.

The ½ mile site used an airbag to support the RCS targets scale models or A-12 section models. A 10-foot diameter, 20-foot high airbag mounted on a table below the lakebed and attached to a gearbox underneath to rotate and tilt the table. A little shed in the pit below the lakebed surface contained the power controls, air compressor, and a sump pump. They called it the swimming pool.

The ¼ mile site lay beneath the lakebed surface and consisted of a 1 square meter fiberglass sphere coated with aluminum tape attached to a fiberglass pole. A gearbox popped up the sphere with the pole tilted toward the radars to minimize having an impact on the sphere RCS and becoming the calibration reference for all RCS measurements. The special projects team's management brought in John Leonardi, a computer expert from National Cash Register to design and build a special purpose computer for radar timing control and data reduction and recording.

As an added feature of Leonardi's computer, he programmed it to recognize what to expect from the systems that it interfaced. The computer programming established standards for what it expected during a mission. If a preflight system evaluation failed to meet the desired signals and levels, it alerted those responsible of the anomaly in for correcting it or knowing it existed and ensured no loss of data simply because someone forgot to flip a switch.

The radars covered frequencies of 500–800 MHz, 1.2–1.5 GHz, and 2.7–3.0 GHz. Each radar contained two antennas pointing down range, one for transmitting and one for receiving. The antennas stacked in front of the radar building ranged from 4 feet above the ground to 30 feet. A dividing plenum of chicken wire erected between the transmitting and receiving antennas reduced transmit pulse into the receiver.

In early 1960, the range designers from SEI with EG&G support built and installed the radars. The SEI personnel consisted of 3 PhDs and a group of senior engineers. John Grace, a senior executive with EG&G, played a significant role in obtaining the various radar systems for Area 51. During this period, the ground range remained fully operational with a four-man group from Lockheed reviewing data and directing EG&G as to which radar systems to use. Additional Lockheed personnel transported and mounted the models at the 1 mile and ½ mile sites.

The RATSCAT, the ground radar range resembled the USAF's RATSCAT built at Holliman AFB on the White Sands lakebed using both SEI and EG&G personnel as consultants.

The chief scientist in the SEI contingent added much more to the projects than radar range designers. He became the countries guru in RCS reduction both for shaping and Radar Absorbing Material (RAM). He introduced RAM-loaded chines on the A-12 to attain an acceptable RCS.

The flight range evolved and articulated with the times, needs, and technology advances. The radars and associated tracking equipment typically used existing systems in the military inventory modified where necessary for the A-12 requirement.

The flight range extended the ground range with the added requirements of tracking the airborne target, measuring it out to 200 nautical miles while recording pitch, roll, yaw and transmit data to the site and calculating RCS returns versus look and depression angle. They measured RCS at 70 MHz for the Knife Rest radar, 172 MHz for the TALLINN (Tall King), and 3.0 GHz for the Fan Song.

The respective radar operators calibrated their radar before each mission. For target tracking and acquisition, the initial design used a 400 MHz telemetry transmitter on the A-12. It used a monopulse tracking receiver on the lip of the big dish to point the dish to a close enough approximation to the target for the 172 MHz or S-band radar to see it.

This higher accuracy track over the monopulse placed the Nike close for acquiring the target for the mission. The special projects team designed and built a bull gear (mechanical parallax correction) to allow switching of antenna control to the correct radar. The team used an X-Band Transponder on the A-12 for a high precision track. However, the 400 MHz monopulse tracking antenna receiver proved unworkable after extensive tests.

The special projects led by engineer Wayne Pendleton performed ground tests by loading an Army pickup truck with a 400 MHz transmitter and some gas-powered generators. They then drove the pickup truck to the hilltops to lock the dish to the transmitter. They never succeeded at grazing angles and concluded that the monopulse array would never work due to ground bounce interference.

For marking the target, the high-power output of the radars accomplished the target measurements. The S-band, a 60-foot dish Bendix SPS-29 radar from Lincoln Labs produced 1 megawatt and produced 750 kWh at 172 MHz. The 4-bay, directional antenna Yagi radar resembled a rooftop terrestrial television antenna or an amateur radio antenna provided a 100-kHz beam at 70 MHz.

The SPS-29, 172 MHz The United States Navy air search radar originally designed for small ships required mounting the output tube in an air cavity filled with a correct dielectric constant oil to lower the operating frequency to 172 MHz. At Area 51, running 3 1/8-inch rigid inch coax up the dish proved challenging as did repairing the broken folded 4-dipole antenna feed. The water-cooled radar transmitter at Area 51 for 425 MHz measurements of target bearing and range required the building of a shack with a swamp cooler to cool the cooling coils. The antenna resembled a bedspring and bore the nickname of bedspring radar.

The special projects team constructed its version of the Russian V-75 SA-2 GUIDELINE A-band, 75 KM Knife Rest radar. It took three tries to successfully build a radar that generated so much ozone the operators could not remain in the closed room during operations.

For measuring pitch, roll, yaw, and transmit data to the site, the telemetry package consisted of a compass and a gyro to generate the parameters. Its electronic converter put the signals into a binary code to modulate the transmitter with a 1 kHz signal phase modulated to produce a pulse for each bit with an error correction remainder code maintaining signal fidelity.

The special projects team originally intended to use an ARC-34 radio. They found it too complicated to operate and switched to using a 400 MHz taxicab radio. The team designed a low observable (LO) antenna about a square foot to match the planes Q-bay contour. Filled with Ecco sorb, it rested 4 inches deep in the camera Q-bay where it transmitted through a crossed slot aperture.

They used an F-101 chase plane for dry runs of the telemetry package. This was because the F-101 weapons bay used a big rotating door designed for launching massive nuclear bombs. It launched the bombs by rotating the door with a bomb attached and executing a pitch up maneuver to lob the bomb away from the F-101. Once he launched the bomb, the pilot rushed the plane out of the bomb kill zone. The special projects team bolted the package frame to the inside of the door and used a standard The United States Air Force UHF antenna attached to the outside of the door.

When installed in the U-2 at Burbank, the package tested OK on a UHF receiver using the UHF LO antenna. When the U-2 arrived at Area 51, the team failed to detect any 400 MHz signals and therefore could not calculate the angles to correlate where the team measured the radar cross-section. This miserable failure prompted a CTJ (Come to Jesus) meeting at Area 51 with severe gnashing of teeth. They disassembled the LO antenna and found it faulty, working as an attenuator instead of an antenna with decent gain and enough radiated energy for a ground check, however, with insufficient gain to achieve any distance. Burbank fixed the problem, and it worked when it arrived back at the test area.

For calculating radar cross-section returns versus look and depression angles, a special purpose computer provided timing for all radar systems. The computer, essentially a DECEMBER PDP-3 prototype recorded the test data on the site. Someone in special projects then hand carried the 1-inch tapes by plane to Waltham, Massachusetts for final calculations before subsequent tests, and for final data reduction.

The ground to slant converter was an arrangement of synchros that acted as a mechanical converter to calculate look and depression angles where the team measured the radar cross-section using pitch, roll, yaw with azimuth, elevation, and range of the tracking radar.

All the radar systems at Groom Lake used an actual square meter sphere for the ground range calibrations. They found this 42-inch diameter sphere too large to throw out of an airplane to measure the radar cross-section in free fall.

Instead, the radar cross-section team used 12-inch diameter world globes coated with aluminum tape for S-band calibrations. For calibrating the 172 MHz radar, they used a resonant dipole (6 feet long) made from a piece of pipe with end slots of fiberglass paddles. The paddles tilted to rotate the dipole as it descended to the ground. The 4X6-inch fiberglass "paddles" used vertical stripes of aluminum tape to provide a target for the Nike tracking radar. The dipole suspended below a parachute with a swivel allowing the dipole to rotate as it descended to the ground.

EG&G used the C-47, flying it as high as it could go without the participants suffering from lack of Oxygen, and threw these targets out the doorway removed for these missions. The Nike acquired the sphere or paddles and tracked them with the bull gear aligning all the antennas on the targets. The plane approached the drop lying about 5 NM out at around 330 degrees in a cross-range heading of 240 degrees. As the 172 MHz dipole rotated, the target returns peaked and fell with the computer calculating the radar cross-section from the peak returns.

As the United States Air Force presence increased, they took over the calibration drop missions using a helicopter for the drops.

They phased out the cardboard world globes for spun cast 12-inch Aluminum spheres and 26-inch spheres for the lower frequencies.

Having reams of data and its reduction proved extremely tough until someone came up with a novel solution to eliminate any clutter contamination. They incorporated an early range gate in the radar, and an operator continually monitored it during a test. When he observed clutter in the early gate (3 miles ahead of the target), he tagged the computer. When the early gate emptied, he tagged the computer again. The team discarded the tagged data for radar cross-section calculations.

The one wavelength in diameter of the inlets on the A-12 presented a primary source for reflections, causing a large radar cross-section at 172 MHz.

Even filling the spikes with RAM required additional effort to reduce their cross-section. Someone proposed generating a plasma cloud in front of each inlet to absorb/deflect radar paints generated by shooting electrons out to the inlets from X-ray guns having a hole in the anode to allow the electrons to exit the x-ray tube. Concerned about the X-rays having an impact on the pilots, the team determined the pilot did not need a lead shield in his seat for his protection from the radiation.

The apparatus filled the Q-bay, and it took a long time to fly an operational unit. The radar cross-section team built a pulsing timer for it to verify radar cross-section reduction. The team learned of the pulsing timer working when the operator of the 172 MHz radar came on the net during the test mission, voicing concern about his receiver oscillating and asking permission to troubleshoot it!

One of the parties involved in the exploitation developed a jammer and brought it to Area 51 to test its operation. It worked by getting in the back lobe of the Fan Song guidance signals where it spoofed the signal.

The simulator on site did not have the guidance antenna/system mounted on the Fan Song antenna pedestal. However, they did have one located approximately 200 feet east of the tracking antennas.

Whenever the A-12 came inbound from the east, the jammer failed to work. After some reflection, the special projects team repeated the test with the A-12 coming inbound from the north. The jammer worked! Apparently, the 200 feet (200 ns in time) caused a fault to the jammer, rendering it unable to recognize the threat. This sort of countermeasure experience proved invaluable for future projects such as the Soviet MiG exploitation project during radar cross-section evaluations.

The Problem

The CIA's first customers at Area 51 came there together because of a common problem. At the end of the 1950s Korean War, USAF pilots flying the North American F-86 Sabre had enjoyed a kill ratio of 10:1 against the MiG planes in MiG Alley. A decade later, the US Air Force, Navy, Marine Corps, and Army found their kill ratio reversed in the Vietnam War.

The ratio increased to 9:1 in the enemy's favor when American pilots encountered the latest Soviet-built MiG-21 Fishbed, the TOP GUN of the Communist world. The United States pilots believed the enemy had superior aircraft. The logical recourse was to acquire and exploit the enemy's air vehicles to discover what made them superior.

Acquiring an Enemy Plane

Craig Thomas was about 30 years off with his USSR technology when he wrote the book about the fictional jet plane in the 1983 movie "Firefox" starring Clint Eastwood. Interestingly, the MiG design produced for the movie closely resembled the stealthy aircraft now appearing in Vietnam. However, that was Hollywood and not the way to acquire enemy planes. In the real world, it was not that easy to obtain an enemy's prized weapon. If you could not capture it on the field of battle, you bought it.

One of the most unusual of the reward campaigns was the attempt to steal a combat-ready Russian MiG-15 Fighter for one hundred thousand dollars during the Korean War.

The Soviets had designed this new fighter just after WWII. The MiG-15, a high-altitude interceptor able to reach almost Mach 1, maneuverable at high-altitude, and armed with cannons, could stay in the air for over 1 hour. The Soviets powered it with their copy of the British Rolls-Royce jet engine that had a higher thrust than the original.

Overall, it performed superior to that of any Western fighter. The MiG-15 outclassed the American P-51 Mustangs, F-80 Shooting Stars, and the F-84 Thunder jets. The Americans waited until December 1950 for the arrival of the swept-wing F-86 Sabre-jet. Even then the MiG-15 climbed faster and remained every bit as maneuverable.

This mysterious plot named Operation Moolah became the Korean War effort to entice a Communist pilot to fly a MiG-15 fighter to an allied airfield for a reward of $100,000.

There are several versions of how this campaign happened. Regardless of the actual scenario, Operation Moolah (GI slang for money) became ready by April of 1953. On 1 April, the UN Joint Psychological Warfare Committee approved Operation Moolah, using data sheets of the Headquarters, 1st Radio Broadcasting & Leaflet Group, 8239 AU, APO 500 dated 20 April 1953. The United Nations Command offered $50,000 to any pilot who flew his MiG-15 to the south, and an additional $50,000 to the first pilot who took advantage of this offer.

The campaign used both radio broadcasts and distribution of aerial leaflets in the Russian, Chinese, and Korean languages. The US believed at the time and later proved that all three countries provided pilots for the air war over North Korea. Each of the three countries (The USSR, China, and North Korea) operated their Air Force units as though fighting against their enemies. No pilot served in another country's Air Force unit. All the MiG-15s, however, carried the North Korean insignia (A red star inside red and blue circles).

In 1953, North Korean pilot, Lieutenant No Kum-Sok, defected with a new Soviet-built MiG-15. He joined a returning flight of F-86s and landing unnoticed at Kimpo Air Base. In doing so, he received $100,000, the bounty offered in the Project Moolah to any pilot delivering a MiG to the United States.

After No had surrendered his aircraft, the USAF took the plane to Okinawa. There, the USAF gave if USAF markings and test-flew it by Capt H. E. "Tom" Collins and Maj Chuck Yeager. The Air Force then shipped the MiG to Wright-Patterson Air Force Base after an unsuccessful attempt to return it to North Korea. It is currently on display at the National Museum of the United States Air Force.

In 1954, Lieutenant No immigrated to the United States where he anglicized his name to "Kenneth H. Rowe." The US evacuated his mother from North Korea to join him in the US. He subsequently graduated from the University of Delaware with degrees in mechanical and electrical engineering. He married an émigré from Kaesong, North Korea, raised two sons and a daughter, and became a US citizen. He became an aeronautical engineer for Grumman, Boeing, Pan Am, General Dynamics, General Motors, General Electric, Lockheed, DuPont, and Westinghouse.

In 1970, No learned from a fellow defector that, as punishment for his defection, North Korea executed his best friend. The North Korean leaders also killed several families and relatives. Those killed also included his battalion commander, vice battalion commander, the battalion's political officer, his regimental commander, and the commanding officer of the North Korean First Air Division. North Korea even killed the air division's chief weapons officer sponsoring No's Communist Party membership.

Lieutenant No's defection was a dream come true for US intelligence who considered this the most valuable technical knowledge of the 1950s because it validated US methods.

Renown test pilot, Chuck Yeager, test flew this aircraft and began a trend that lasted nearly forty years. Within three years, no less than four Polish pilots flew their MiG-15s to freedom. The planes obtained through defections enabled the USAF to test fly the MiG-15 at Wright-Patterson, Eglin, and PAX River against the B-47, B-36, and F-86.

The success of the MiG 15 exploitations sparked a continuing quest for possession of enemy planes. In virtually hundreds of cases, warring nations have offered cash to the enemy. Sometimes the countries paid for defections or weapons. Other times governments used cash for aid to friendly personnel or to buy loyalty to a friendly government. During the Vietnam War, the US offered gold to North Vietnamese soldiers and Viet Cong Guerrillas. The gold persuaded the enemy to aid American pilots forced down over Communist-held territory.

Early in the Vietnam War, US Aircraft dropped leaflets signed by the Commander, US 2nd Air Division. The leaflets offered $35,000 in Vietnamese currency for the safe return of Lieutenant George E. Flynn shot down while flying an A1E Skyraider over the Dong Thai outpost, Hieu Le District, Kien Giang Province.

1953, North Korean pilot, Lieutenant No Kum-Sok

MiG-15

Secretly Exploiting a Foe's Aircraft

Exploiting a potential enemy's aircraft in secret is not something new. In the early 1950s, the USAF tested a Yakovlev Yak-23 at Wright-Patterson Air Force Base, Ohio. After exploiting the plane, the Air Force loaded the fighter into a C-124 Globemaster and returned it to a cooperative owner in Eastern Europe. The tests of this Yak-23 remained secret for more than 40 years after they took place.

The Soviet VVS had an opportunity to study a pair of US F-14 Tomcats and extensively test them against current Soviet fighters, this occurring in the middle of a desert just outside of Moscow-at the Zhukovsky Flight Test Center. Most knew the center as "Ramenskoye" airbase because of the nearby Ramenskoye highway. The Tomcats came from Iran.

Another Russian organization, the Central Aerohydrodynamic Institute, known by its Russian abbreviation "TsAGI" consistently participated in testing of foreign aircraft and their components. At this facility, the Russians tested two F-86 Sabres captured by the VVS in Korea. They tested numerous other foreign aircraft obtained during various local conflicts, during the Korean and Vietnam wars, as well as reconnaissance aircraft shot down over or near the USSR. The Moscow Aviation Institute or MAI frequently participated in this research.

Soviet-built Mikoyan-Gurevich MiG-21, the most advanced Soviet fighter plane at the time, began production in 1959. Egypt, Syria, and Iraq received numerous aircraft, however, not Israel. The Israeli Mossad undertook Operation Diamond (Hebrew: Mivtza Yahalom) to acquire a MiG-21. The operation began in mid-1963 and ended on August 16, 1966, when an Iraqi Air Force MiG-21, flown by a defector, landed at an air base in Israel.

In 1964, an Iraqi-born Jew, Yusuf, had contacted Israeli services in Tehran (Iran and Israel had diplomatic relations at the time) and Western Europe. At ten years, old, Yusuf worked as a servant for a Maronite Christian family. His girlfriend's friend married an Iraqi pilot named Munir Redfa. It annoyed Redfa that his Christian roots prevented his promotion in the military. Yusuf understood that Redfa felt annoyed about orders to attack Iraqi Kurds and wanted to leave Iraq.

A female Mossad agent befriended Redfa. He told her about Iraqis forcing him to live far away from his family in Baghdad. He also claimed his commanders did not trust him and allowed him to fly only with small fuel tanks because of his Christianity.

Redfa traveled to Europe to meet with Israeli agents who offered him $1 million, Israeli citizenship, and full-time employment. Israel sent several Mossad agents to Iraq to assist the transfer of Redfa's spouse Betty; their two children aged three and five, his parents, and some other family members out of the country. Betty and their two children went to Paris for what she thought to be a summer vacation. They took the other family members to the Iranian border. There, Kurdish guerrillas helped them to cross into Iran, and then to Israel.

The opportunity to defect came about on 16 August 1966. Jordan radar detected Redfa flying over the northern Jordan. They stood down when Syria assured them of a plane belonging to the Syrian Air Force and flying on a training mission. Two Israeli Air Force Dassault Mirage IIIs met Redfa's plane as it entered Israel, and escorted him to a landing at Hatzor, an Israeli Air Force base located in central Israel near kibbutz Hatzor. Later, at a press conference, Redfa said that he had landed the plane on "the last drop of fuel."

Soon after his defection, the Air Force changed the numbering on Redfa's MiG to 007, reflecting the way it arrived. The Israelis analyzed the jet's strengths and weaknesses against IAF fighters. On May 1967, director of the CIA Richard Helms noted that Israel had proven that it had made proper use of the aircraft. On April 7, 1967, during aerial battles over the Golan Heights, the Israeli Air Force brought down 6 Syrian MiG-21s without losing any of its Dassault Mirage IIIs.

In January 1968, Israel lent the MiG to the United States, which evaluated the jet under the HAVE DOUGHNUT program. The transfer helped pave the way for the Israeli acquisition of the F-4 Phantom, which the Americans felt reluctant to sell to Israel.

The Air Force Systems Command had large divisions that handled different aspects of technology: Aeronautical Systems Division, Electronic Systems Division, Armament Division, and others developed new technology for the AF. Foreign Technology Division was (and still is, as NASIC) involved with analyzing foreign military hardware, particularly foreign aircraft, some air defense systems, and space systems. They figured out how foreign fighter jets worked based on any info they could get: signals from the fighters, radar tracks of the fighters, photographs, maybe human sources, whatever they could get. If the engineers could get their hands on real hardware, that would be ideal to fill in gaps in knowledge and confirm things they were able to piece together. In 1968, FTD was part of the Intelligence Community, as NASIC is today. Intelligence organizations had an intelligence community 'chain of command,' separate from their military chain of command. The CIA director was Director of Central Intelligence, and the CIA was only one small part of the community under his (loose) control. Beneath him was a Director of Military Intelligence, who ran the Defense Intelligence Agency, and somewhat controlled all intelligence activities in the DoD. There were service intelligence chiefs as well. Decisions about the production of intelligence reports were handled somewhat through the intelligence chain amidst lots of cooperation (and rivalry) among the military services and civilian agencies. Once the intelligence community was involved, there was an honest effort to get all the interested players involved.

Names of personnel involved remain secret to protect their privacy; however, declassification has authorized personnel involved in these specific projects to discuss data within the reports released by the various governmental agencies conducting the MiG projects.

A relatively recent photo of a MiG-21 in an Israeli museum painted to look the way the HAVE DOUGHNUT aircraft looked after it's return to Israel in 1968 and before it was

Chapter 3 - MiG-21 Project HAVE DOUGHNUT

Loss of American Air Superiority.

The Air Force and Navy came to Area 51 suffering from the loss of American air superiority. The problem was new and unexpected. With the problem being universal to all American military branches made it appear to be the enemy having superior planes. However, then again, the F-4 was getting whacked by the Korean War vintage MiG-17 Fresco. How could that be?

At the end of the Korean War in the 1950s, USAF pilots of the North American F-86 Sabre alone enjoyed a kill ratio of 10:1. The US aerial warfare changed in the next war. The United States went into the Vietnam War against an enemy still shooting cannons at US planes armed with only air-to-air missiles.

US pilots entered the Vietnam War with advanced weapons, and aircraft against a peasant Army and an Air Force armed only with Korean War vintage weapons. They faced other problems as well. American troops could not consistently track low flying MiG on the radar. Restrictive rules of engagement (ROE) hampered the US pilots more than anything else. The White House required pilots to visually acquire their targets, nullifying much of the advantage of radar guided missiles, missiles that proved unreliable when used in combat.

The new generation of United States fighters brought the development of air-to-air missiles, fighter aircraft, such as the US Navy's F4H Phantom II, later redesignated the F-4 in 1962. The United States entered the war with fighter aircraft designed from the start with only air-to-air missiles. US planes carried both radar-guided AIM-7 Sparrow III and the shorter-range AIM-9 Sidewinder infrared-guided missiles. The United States entered the Vietnam War with new missiles and a new attitude of dog-fighting now obsolete.

New Navy and Marine Corps F-4 crews received extremely limited air-to-air training. They only flew about ten flights and received little useful information. By 1964, few remained in the Navy and Marine Corps to carry on the tradition of classic dogfighting. Then came the Vietnam War.

The early years of the air war over North Vietnam showed the erroneous faith placed in missiles. Between 1965 and the bombing halt in 1968, the USAF had only a 2.15:1 kill ratio. The US Navy fared slightly better with a 2.75:1 rate. The Navy lost an American F-4 Phantom II, F-105 Thunderchief, or F-8 Crusader for roughly every two North Vietnamese MiG-17 Frescos or MiG-21 Fishbed planes shot down.

The continuously growing percentage of United States fighters lost in air-to-air combat proved worse than losing the 10–1 kill ratio enjoyed by the US aircrews of the Korean War. The 1966 3 percent loss of US aircraft losses due to MiG rose to 8 percent in 1967 and then climbed to 22 percent for the first three months of 1968.

The emphasis on air-to-air missile interception left the fighter combat crews with only the sketchiest knowledge of dog-fighting.

Originally conceived as a naval fleet air defense aircraft, and later adapted as an Air Force fighter-bomber, the design of the F-4 made it ill-suited for a tight-turning dogfight. In contrast to the lighter MiG-17, the large and heavy F-4 lost energy and airspeed when in turn. However, the MiG-17's superior turning capability allowed it to close to the gun range where all too often, hits from the MiG-17's "outmoded" cannons destroyed the F-4.

The Soviet sponsors and North Vietnamese Air Force commanders knew the MiG-21 Fishbed's limitations and planned around them. They committed their fighters only if seeing a good chance of success and subsequent escape. During that stage of the Vietnam War, 80 percent of the North

Vietnamese Air Force kills occurred with the victims unaware of their even coming under attack.

"Red Baron" tactics, initiating an attack from a cross-course intercept accounted for much of their success. MiG-21 fighters factored by ground control intercept radar from Chinese airspace positioned behind the F-4 Phantoms bombing targets north of Hanoi. Known as "blow-through," the MiG launched ATOLL missiles at F-4s pulling up from their target. The MiG then zoomed back to a political sanctuary in China.

US fighter design followed the success of the small, highly maneuverable F-86-day fighter in the Korean War by changing to emphasize maximum speed, altitude, and radar capability. Now, fighting in a new war, the fighters found the chances since the Korean War came at the expense of maneuverability and, pilot vision. US pilots no longer possessed the attributes needed for close combat. During the latter part of the Vietnam War, this trend reached its extremity in the McDonnell Douglas F-4 Phantom, the dominant fighter for both the US Air Force and Navy.

The F-4 interceptor design met that required by the fleet defense mission using rapid climb to high-altitude, high supersonic speed, and radar-guided missiles to shoot down threat aircraft at a long distance. The F-4, though originally an interceptor for defense of the fleet against air attack — never executed such a flight. No US fleet had ever come under air attack since the beginning of the jet age.

Combat Losses in SEA (South East Asia)

The Air Force and Navy came to the CIA at Area 51 at the height of the Vietnam War. Technologically advanced American aircraft from both the US Navy and the USAF filled the skies over Vietnam. The air battles lacked the past glories in the 1950s skies over Korea's MiG Alley. Their aircrews faced carefully trained North Vietnamese pilots and skilled warriors such as top ace, Nguyen Van Coc; credited with seven aircraft and two Firebee unmanned aerial vehicles destroyed. His aircraft victories included two Air Force F-4s, one Navy F-4B, two "Wild Weasel" F-105s, one F-105D, and the only F-102A kill of the war.

The North Vietnamese Air Force's first jet air-to-air engagement with US aircraft occurred on April 3, 1965, with the NVAF claiming the shooting down of two US Navy F-8 Crusaders. Though never confirmed by US sources, the Navy acknowledged having encountered MiGs. Consequently, April 3 became "North Vietnamese Air Force Day."

On April 4, 1965, the VPAF (NVAF) scored the first confirmed victories acknowledged by both sides. It shocked the US fighter community when relatively slow, post-Korean era MiG-17F fighters shot down advanced F-105 Thunderchief fighter-bomber aircraft attacking the Thanh Hóa Bridge. The two downed F-105s carried their normal heavy bomb load, making them unable to react to their attackers.

Also in 1965, the USSR supplied NVAF with supersonic MiG-21s, which NVAF used for high-speed GCI controlled hit and run intercepts against American air strike groups.

By late 1966, the MiG-21 tactics became so effective that the US mounted an operation to deal with the MiG-21 threat especially.

Led by Colonel Robin Olds on 2 January 1967, Operation Bolo lured MiG-21s into the air to intercept what appeared as an F-105 strike group. Instead, they found a sky full of missiles armed F-4 Phantom IIs set for aerial combat. The VPAF lost almost half its inventory of MiG-21 interceptors, at the cost of no US losses. The VPAF (NVAF) stood down for additional training after this setback.

Using the F-4 in Vietnam as a fighter rather than as an interceptor severely miscast the plane at high expense to the Air Force and Navy aircrews. The air-to-air kill ratio sometimes dropped as low as 2:1 against inferior North Vietnamese pilots flying small, highly maneuverable MiG-21s. The Navy and Air Force knew something was wrong when comparing this to the lower kill ratio to the 13:1 ration in Korea

Politics Part of the Problem

The first exploitation of a Soviet-built MiG-21F-13 (Fishbed-E) fighter-interceptor occurred during a troubling time in US history.

The conflict between a conservative regime and a growing number of anti-Vietnam protesters spreading from the campuses to the streets and federal agencies secretly investigated the loyalty of American citizens.

It did not help the US air war when the Johnson administration forbids targeting North Vietnamese airfields, parked aircraft, command centers, and main radar installations.

The US entered the Vietnam War with a move away from cannon fire to air-to-air missiles. The US found the dogfight alive and well in the Vietnam War and missiles not yet ready to replace the gun. The Top Vietnamese ace of the Vietnam War claimed nine kills: seven manned aircraft and two UAVs. In just one day, in December 1966 the MiG-21 pilots of the 921st FR downed 14 F-105s without any losses. Another change, the much more extensive use of helicopters and spotter planes provided easy targets for the enemy's MiG fighter planes.

The F-100 came close in the air superiority role in conventional warfare but arrived too soon for the M-61, therefore, arriving equipped with four M-39s. Only the F-104 and F-105 came fitted with the M-61 Gatling gun. The Air Force brought the F-4 into its inventory for the air superiority role without a weapon. Because of the Vietnam War losses, the fighters employed the M-61 Gatling gun carried externally in the SUU-16 pod.

The AIM-7, designed as an anti-bomber weapon, lacked the broad range for firing or maneuverability needed in a fighter-versus-fighter engagement. However, the AIM-9 kill rated somewhat better, about twenty percent during the latter part of the 1965–1968 campaign. The Air Force F-4E with an internal gun made its debut in the war in 1968 following what the Air Force and Navy learned at Area 51 during Project HAVE DOUGHNUT.

The Vietnam War conflict saw the first broad-scale tactical deployment of helicopters. The enemy shot down over 8,000 helicopters, costing the US over 5,000 US helicopter pilots.

The US lost 2,200 Bell UH-1 Iroquois (Huey) helicopter pilots in the war. This number of losses does not count all the Cobras, CH-53, Chinooks, H21 Shawnees, Sioux, Choctaws, Sea Stallions, Jolly, and other choppers used in the war. Nor did this count the OH-58 Kiowa, OH-6 Cayuse, AH-1 Green Giants, CH46s, etc., or the other crew aboard those shot down helicopters, men other than pilots, just members of the chopper crew, door gunners, crew chiefs, etc.

Not counting helicopter pilots and aircrew, the US lost over 6,000 fixed-wing/propeller/jet US pilots and aircrew, killed, or missing during the Vietnam War. The USAF lost about 2,584 men, and the USN lost about 2,555 men. The USAF and USN together lost well over 2,000 fixed-wing aircraft along with the US Army losing the 8,000 rotor-wing aircraft

Alternatively, Was it the Planes?

Again, the US needed to acquire a MiG to find out the cause of its losses in the air war. The MiG-21F-13 mystery plane no longer appeared as a mystery once it first flew over Area 51 in 1968 with a YF-110B designation.

The HAVE DOUGHNUT story begins on 16 August 1966 in the Middle East a year after Operation Rolling Thunder, the sustained air bombardment of North Vietnam began.

Monir Radfa, an Iraqi Air Force captain, took off in his Mikoyan MiG-21F-13 from Rashid Air Base outside of the Iraqi capital, for a local navigation exercise. Instead, he dashed the southwest at low-level, intending to defect. The Jordanian RJAF's Hawker Hunters, too slow at low-level, failed to intercept him as he streaked low across their country. He entered Israel, lowered his landing gear, and wagged his wings to two intercepting Israeli Mirage III fighters to signal his intentions. The Israelis escorted him to Hatzor Air Base, where they gave him asylum.

The single-seat, supersonic, single-engine, delta wing, sweptback tail MiG-21 Fishbed fighter jet threatened Israel as one of the most potent fighters in the Arab Air Forces. The Israelis immediately set about flight-testing the mid-wing monoplane for over 100 hours. Over the next 12 months, the Israelis learned its strengths and weaknesses and taught the Mirage III pilots how to defeat the MiG in a dogfight.

Israel initially hesitated to share its prize with the United States. However, Israel eventually did so after agreeing brokered by the US Defense Intelligence Agency (DIA) to lend the MiG-21 to the US for study. In exchange, the US allowed them to buy the F-4 Phantom II, the American front-line fighter of the day.

The exchange followed a time when the Israelis made several overtures to the Johnson Administration to purchase the Phantom only to have President Johnson rebuff them out of a fear of escalating matters in the Middle East. Having the MiG-21 provided the Israelis enough advantage to get the Phantoms headed their way and the US finally an able close study of its vaunted adversary in the skies of Vietnam.

The Air Force Foreign Technology Division

The Air Force Foreign Technology Division (FTD) evolved from the Air Technical Intelligence Center (ATIC) established on May 21, 1951. ATIC analyzed engine parts, and the tail section of the Korean War Mikoyan-Gurevich MiG-15 obtained during Project Moolah. FTD evolved as a USAF field activity of the assistant chief of staff for Intelligence under the direct command of the Air Materiel Control Department.

The Foreign Technology Division, (FTD) formed in 1961 at Wright-Patterson Air Force Base in Dayton, Ohio analyzed foreign military hardware, particularly foreign aircraft, some air defense systems, and space systems. It started out a component of the Air Force Systems Command (AFSC). Large divisions within the Air Force Systems Command handled different aspects of technology: Aeronautical Systems Division, Electronic Systems Division, Armament Division, and others developed new technology for the Air Force. FTD provided the National Security Council with intelligence estimates through the 1962 United States Intelligence Board at the CIA.

In 1968, FTD played a part of the Intelligence Community, as NASIC is today. Intelligence organizations operated with an intelligence community 'chain of command,' separate from their military chain of command. In early 1968, FTD delivered the HAVE DOUGHNUT MiG-21 to Area 51 for testing and evaluation of foreign aircraft technology, this becoming the longest continuing United States classified military aircraft program ever.

Much of the preparation for exploitation of the MiG-21 HAVE DOUGHNUT vehicle occurred in the Flight Dynamics Laboratory at Wright-Patterson. (The author transferred from the Seven Sisters' Beatty radar site to the Flight Dynamics Laboratory in preparation for HAVE DOUGHNUT. He returned from Wright-Patterson and reported in with the Special Project team at Area 51 for the arrival of the HAVE DOUGHNUT MiG.)

The focus of Air Force Systems Command limited the use of the fighter as a tool with which to train the front line tactical fighter pilots. Air Force Systems Command recruited its pilots from the Air Force Flight Test Center at Edwards Air Force Base, California, who were usually graduates from various test pilot schools. Tactical Air Command selected its pilots primarily from the ranks of the Weapons School graduates.

CENTRAL INTELLIGENCE AGENCY
WASHINGTON, D.C. 20505

EO 12958 6.1(c)>25Yrs

APPROVED FOR RELEASE
DATE: MAY 2002

25 July 1969

MEMORANDUM FOR: The Director of Central Intelligence

SUBJECT : Acquisition of Soviet Technical
Manuals for the MIG-21PFM Aircraft

1. I wish to inform you that the Clandestine Service
has just acquired three Soviet technical manuals for
the MIG-21PFM [FISHBED F] interceptor. These manuals
are the first documentary data to be acquired on this
most recently deployed version of the MIG-21. The manuals
cover performance characteristics, tactics, and maintenance.

2. The MIG-21PFM is a further development and improvement of the MIG-21FL (modified FISHBED D), for which the
Clandestine Service obtained manuals previously. The
MIG-21PFM is now the main production model of this widely
deployed fighter, which is operational or becoming
operational in North Vietnam, North Korea, Cuba, Afghanistan,
India, the UAR, and other countries. This version is now
the most modern interceptor available to the air defense
forces of the Warsaw Pact countries, excepting the USSR.

3. The most useful manual is expected to be the
Combat Employment manual, which updates tactics used by
the modernized MIG-21 and which probably is the primary
combat guidance for North Vietnamese MIG-21 pilots.

4. These manuals contain confirmatory and new data
on the RS-2US (ALKALI) air-to-air missile, a booster
starter system to permit rocket-assist take-off, and the
first documentary information available on a Soviet boundary
layer control system (SPS) for an aircraft. The latter
two systems (RATO and SPS), not previously incorporated
together into an operationally deployed Soviet aircraft,
permit the MIG-21PFM to operate from poor quality runways
and improvised airfields.

NLNP
Mandatory Review
Case NLN 99-
Doc. 15

EO 12958 3.3(a)>25Yrs

SECRET

-2-

EO 12958 6.1(c)>25Yrs

EO 12958 3.4(b)(1)>25Yrs
(b)

5. Analytical elements of the Intelligence Community are being notified via normal distribution channels of the acquisition of the above documents and are being provided pertinent extracts from the documents in _____. The FBIS is currently translating the documents and the translations will be disseminated in the CSDB series to all pertinent consumers.

EO 12958 6.1(c)>25Yrs

Thomas H. Karamessines
Deputy Director for Plans

EO 12958 6.1(c)>25Yrs

47

S E C R E T

Distribution: The Director of Central Intelligence

The Assistant to the President for
  National Security Affairs

The Director of Intelligence and Research
  Department of State

The Director, Defense Intelligence Agency

The Director, National Security Agency

Director, National Indications Center

The Deputy Director of Central Intelligence

Deputy Director for Science and Technology

Deputy Director for Intelligence

Director for National Estimates

Director for Current Intelligence

Director for Strategic Research

Director for Economic Research

Director for Scientific Intelligence

*The Mystery Jet – the MiG-21F-
13 Fishbed.*

The HAVE DOUGHNUT MiG-21F-13 delivered to the Israeli Defense Forces Air Force and lent to the US for exploitation. Air Forces' Foreign Technology Division managed the exploitation in a joint Air Force/Navy technical and tactical evaluation at Area 51 with delivery on January 23, 1968.

HAVE DOUGHNUT Exploitation Team Leaders

Major Fred J. Cuthill, chief of the Air Force Flight Center Branch, served as the AFSC project officer to evaluate performance and handling qualities of the HAVE DOUGHNUT MiG-21.

Project officer Comdr. Thomas J. Cassidy, Jr., Naval Air Test and Evaluation Squadron 4 (VX-4) along with Lt Col Joe B. Jordan as the TAC project officer for the Air Force TAC conducted tactical evaluation of the aircraft. (VX4 did not own any aircraft but flew with and were trained by the Red Hats. VX4 pilots later flew with the Red Eagles as well.)

Later known as the Red Hats, the rest of the customer's exploitation team consisted of Robert G. Ashcraft *–TAC, Gerald D. Larson – TAC (1137th SAS), William T. "Ted" Twinting – AFSC. *In December 1977, the Air Force activated the 6513th Test Squadron Red Hats at Edwards AFB to support evaluation of foreign aircraft. Functional detail of the Red Hats remains classified. The Air Force deactivated the Red Hats in October 1992 and immediately reactivated the group as the 413th Flight Test Squadron, providing test and evaluation capability for electronic warfare (EW) systems.

Technical Evaluation of the MiG-21

Project Background
- The US borrowed a Soviet-built MiG-21F Fishbed from 23 January to 8 April 1968
- The Foreign Technology Division of AFSC let the exploitation using expertise from AFFTC, ASD, TAC, and NWC.
- The purpose substantiated and supplemented existing threat data.
- Included ground and flight testing.
- 102 flights (77 flying hours) in 30 days of flying.
- The US gave the jet back when they finished with it.

This recently declassified project reveals how it prompted the genesis of the Navy's TOP GUN Weapons School, Air Force's RED FLAG exercises. Out of it came the recently declassified USAF Red Eagles CONSTANT PEG program, the Red Hats (still classified), and other similar MiG exploitation programs, the details of which remain classified.

The public release of HAVE DOUGHNUT, HAVE DRILL, and HAVE FERRY information is limited strictly to the referenced reports in this book. To protect international agreements and national security, the author specifically excludes any technical details about any subsequent exploitation efforts. To the extent authorized, the author discusses only the declassified exploitation programs that continued another 20 years as follow-on programs generated by the successes of HAVE DOUGHNUT, HAVE DRILL, and HAVE FERRY.

The Foreign Technology Division of Air Force Systems Command, a predecessor of today's Air Force Materiel Command took responsibility of the highly classified tests. The word `HAVE' in project names such as HAVE DOUGHNUT (MiG-21) or HAVE BLUE (the prototype for the F-117 stealth fighter), identifies the project as belonging to FTD, now known as AFS.

The Air Force disassembled and transported the project HAVE DOUGHNUT MiG by a Lockheed C-5 Galaxy to the USAF's secret flight test facility at Groom Lake, Nevada. In Hangar 5, assembly began the next day with FTD personnel from Wright-Patterson AFB in Ohio inspecting, adjusting, repairing, and operationally checking all systems.

The HAVE DOUGHNUT program called for two categories of flight-testing of the MiG. The first type concerned technical analysis-performance, flight envelope, engineering, structures, and so on. For this, the Air Force prepared at Wright-Patterson Air Force Base by developing a "black box" and acquiring instruments to record their test results.

The CIA at Area 51 prepares for the testing by ordering an X-band Nike radar system from Fort Bliss, Texas. At the same time, the author, Nike-trained Thornton D. "TD" Barnes transferred from the NASA X-15 High Range in Nevada to the Flight Dynamics Laboratory at Wright-Patterson. There, he helped develop and test instrumentation for the impending MiG exploitation at Area 51. Barnes, an Area 51 Project OXCART participant in three separate stages of that program, knew what the MiG program needed to interface with the equipment at Area 51.

At Wright-Patterson, Barnes used his performing integrity tests on the Apollo 1 space capsule as his cover. He returned to Nevada for assignment to the special projects exploitation team at Area 51.

He arrived at about the same time the CIA negotiated with the US Army at Fort Bliss, Texas for an X-band Nike Hercules radar system used by Barnes at the McGregor Missile Range in New Mexico.

Barnes' deployment to the McGregor Range followed two years electronics schooling on the Nike Ajax, Nike Hercules, and HAWK missile and radar systems while serving in the Army.

The CIA acquisition of this Nike radar system occurred long after the closure of Project OXCART at Area 51, thus, identifying it as a replacement for the older Ajax system and intended for direct participation in the MiG exploitation projects.

Wayne Pendleton, who served in the Army at Fort Bliss at the same time as Barnes drove to Fort Bliss with a truck tractor to tow the radar van back to Groom Lake. While Barnes attended 2 and a half years of formal schooling in missile and radar electronics, Pendleton, a graduate from WPI with an EE degree served as a radar and countermeasures equipment repair officer while at Fort Bliss.

The planned exploitation of the MiG-21 involved exploitation during the technical phase. Following the technical phase, the project entered a choreographed tactical exploitation phase where the MiG engaged in mock dogfights and comparison flights against US fighter aircraft. The AFSC/FTD emphasized the technical exploitation of the MiG. Consequently, most of the HAVE DOUGHNUT flying concentrated on technical analysis.

The special projects test headquarters became a constant rotation of customers working with the special projects team providing mission control rooms, radar and radio coverage, and data processing and storage. The MiG-21 Fishbed remained equally busy, flying a mission against one group, returning to refuel, and heading out to fly against another of those having an aircraft to compare with the MiG.

Each mission for the MiG-21 meant a mission for Barnes and the other radar operators. Barnes' X-band Nike performed the tracking of the MiG-21, requiring him to track every mission. Except for RCS evaluations, the remainder of the radar systems stood down. Data acquisition and storage likewise participated in virtually every mission. The sounds of aircraft taking off and returning never ceased until the last participating planes landed for the day.

During the HAVE DOUGHNUT tactical evaluations of the MiG-21, the Air Force Tactical Air Command (TAC) jointly participated with the US Navy and other government agencies. As a team, they analyzed the MiG-21F-13 Fishbed-E day fighter weapons systems. The evaluation included the performance of the MiG-21F-13 against the ECM systems of the B-58 and B-52. Jointly, they determined its tactical capabilities as a weapons system against their respective warplanes in air-to-air environments.

The USAF FTD conducted acoustic measurements, Photometric coverage, engine flight test data, and along with the Naval Air Propulsion Test Center and AFFTC, exploited the propulsion system. The Naval Air Propulsion Test Center conducted static, transient ground tests, and the Naval Missile Center evaluated the SIRENA Tail Warning Receiver, RSIU-3M VHF transceiver, and SRO-2 IFF equipment. The results of the technical exploitation included flight test and performance data, maintenance summary, system and subsystem characteristics, design features, and technological information.

Both the Air Force and the Navy recognized the MiG-21 as a day fighter at medium and low altitude. They also knew it used intercept tactics where the missions called for GCI vectoring the MiG into the rear hemisphere where they attacked in high-speed, single-pass attacks. Due to tactics of US aircraft deploying in SEA, Southeast Asia, the Air Force, and Navy seldom encountered the MiG-21 in the high-altitude, point intercept role. The exploitation at Groom Lake provided both the opportunity to evaluate the MiG planes in that environment as well in an air-to-ground environment.

25X1A2g

█████-9224-68

Copy 6 of 8

21 AUG 1968

MEMORANDUM FOR: Chief, Applied Physics Division, Office
of Research and Development

ATTENTION: █████████████████ 25X1A9a

SUBJECT: Nike Hercules Radar

REFERENCE: a. Memo for Record from AD/M/OSA; dated
28 May 1968; Subject: Arrangements
for Shipment of Nike Hercules Radar
25X1A2g ████-9045-68)

b. Memo for Record from D/M/OSA; dated
17 June 1968; Subject: Shipment of
Nike-Hercules Radar" ████-9106-68) 25X1A2g

1. Reference B, Paragraph 4, states that ORD had
budgeted for spare parts for the Nike Hercules radar.
The M & S van, part of the radar system, contained a year's
supply of spares for field support costing $53,850.72.
The M & S van, including the spares, was assumed to be part
and parcel of the radar system transferred at no cost to
the Agency.

2. 25X9A5 Upon the movement of the radar system, Fort Bliss
(Capt. ██████████████████████████ insisted that
the spares in the M & S van were to be transferred at cost
to the Agency. This was later confirmed by Lt. Col. F. D.
Burnett, DSC/LOG, Department of Army.

3. 25X9A5 The spare parts from the M & S van were returned
to Capt. ██████████████████████ on 7 August 1968
and an automatic receipt document received and is filed with
D/M/OSA.

25X1A9a

Colonel   USAF
Deputy for Materiel, OSA

25X1A2g

Handle via
█████████

The First Exploitation Project at Area 51.

The first exploitation at Area 51 involved the HAVE DOUGHNUT, an export MiG-21F-13 (Article 74) with an aircraft manufacture date in the last quarter, 1963. The aircraft had approximately 135 hours on it—the engine 165 hours. No ATOLL missiles came with the deal, so the exploitation team substituted almost identical AIM-9B Sidewinder missiles.

An R-37F axial flow turbojet with 12,650 pounds max afterburner thrust powered the clipped delta wing planform with sweeping tail surfaces.

The MiG-21 Fishbed-E weighed 11,017 pounds empty, 17,286 pounds at takeoff, and a maximum of 18,072 pounds. It had a wingspan of 23.47 feet, length (without pitot boom) of 44.2 feet, and a height of 13.5 feet. Its armament included one 30 mm cannon with 60 round capacity, two ATOLL missiles, and a total bomb load on all three stations of 3,300 pounds. Its maximum performance speed was 2.05 Mach with a 57,500-foot service ceiling and a strike radius of 370 nautical miles with external fuel. It carried a load of 4,600 pounds internal and 880 pounds in the centerline tank.

Test Environment and Procedures

The HAVE DOUGHNUT participants planned the technical exploitation of the HAVE DOUGHNUT MiG at Wright-Patterson. Barnes, a veteran of Project OXCART knew the equipment available at Area 51. Utilizing this knowledge, they determined what to seek and how to obtain it using what one might describe as equipment that included a specially designed black box to extract the needed information.

AFFTC Performance Evaluation
Performance Sorties   17
Stability and Control Sorties   9

Site-installed Instrumentation
Once the plane arrived at Hangar five at Groom Lake, the exploitation team installed the instrumentation necessary for the exploitation. Including:
- An oscillograph, 12 channels-nav light switch/cannon switch,
- Gyros-Pitch, Roll, Yaw plus rates–vertical tail,
- Fuel Flow Meters-total and normal,
- Photo Panel-Airspeed, Altitude, Mach, Free Air Temp, & (in nose) Clock,
- Instrument panel-A-13 clock, Airspeed, Altimeter, Accelerometer, Stopwatch, Engine fuel temp, two Triad 16 mm cameras, voice recording system
- Cockpit-two Triad 16 mm cameras,
- Voice recording system,

- Battery, and
- A UHF radio.

This was the first time for most of the exploitation team to have physical possession of the MiG-21. The two exploitation teams, the Naval Weapons Laboratory, and a contractor for NWC (Naval Weapons Center) made a detailed structural study of the project aircraft. The Falcon Research and Development Company, under contract to NWC, examined the plane. When it came to construction, the company found it in keeping with recognized shoddy Soviet design philosophies. However, other systems indicated a healthy state of Soviet technology.

A serious, yet almost humorous incident early in the program validated the warranted concern for the unexpected. The team hooked the plane to its laboratory instrumentation on the day of the first engine run-up of the MiG-21. Virtually everyone related to the project came to witness the event. Someone noticed everyone wearing their security badges with attached dosimeters and film badges and suggested gathering up all the badges to prevent accidentally sucking one of them into the engine during the run-up.

The exploitation team gathered up all the badges and gave them to one of the military spectators to hold during the test. With the MiG engine screaming at military power and the plane's brakes straining to hold the aircraft in place, everyone gravitated to the front of the plane—including the one holding all the badges. Instead of sucking a badge through the engine, the engine sucked ALL the badges though.

The impact of the security badges with radiation dosimeters attached, of course, caused severe damage to the impeller blades of a relatively new engine (165 hours) that had yet to fly at Area 51. Fortunately, some of the Pratt & Whitney J58 engine engineers remaining on site from Project OXCART repaired the engine enough to fly the HAVE DOUGHNUT tactical flights.

Missile configuration

- ■Non-firing AIM-9B used to replace ATOLLS
- ■AIM-9 rail with laminated plywood/fiberglass and steel fittings
- ■ No performance change except slight improvement at low speed

The MiG-21 in Hangar Five at Groom Lake. Note the US Air Force decals designated as a YF-110B

Calibrated test instruments installed for exploitation

Airspeed Dial Face

Mach Meter Dial Face

Rate of Climb

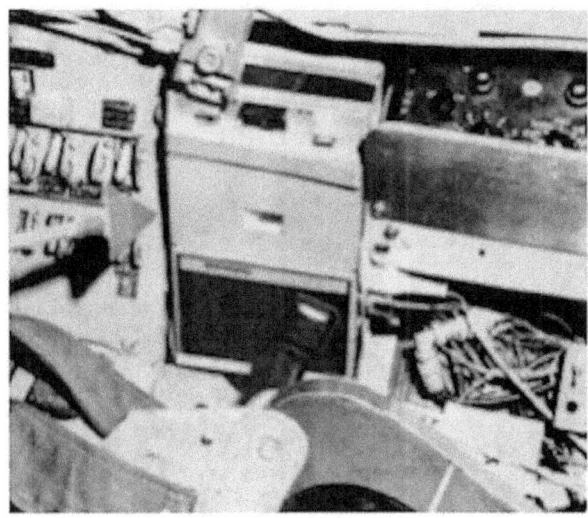

Voice Recorder

MiG-21-F-13 Fishbed cockpit instrument panel

⊞

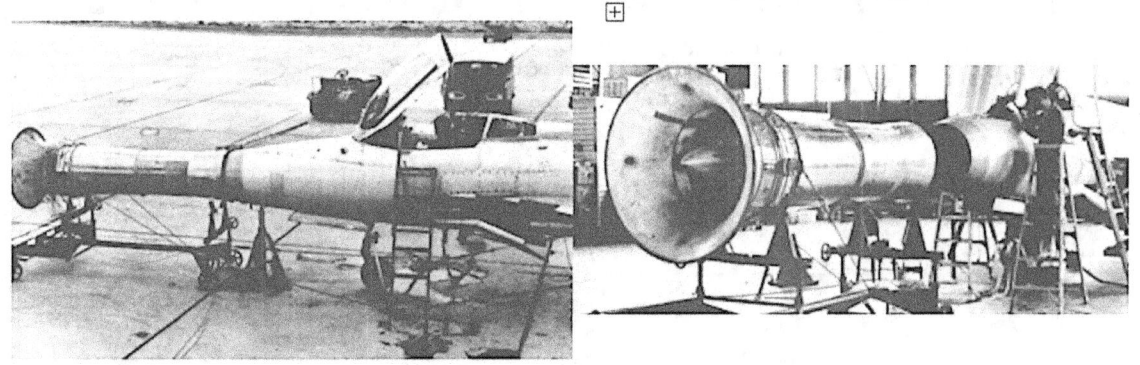

Bellmouth installation attached to aircraft during engine airflow evaluation.

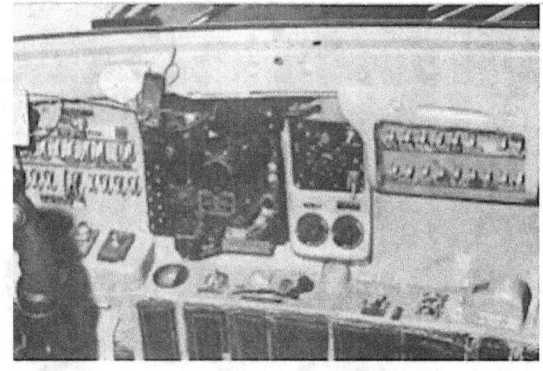
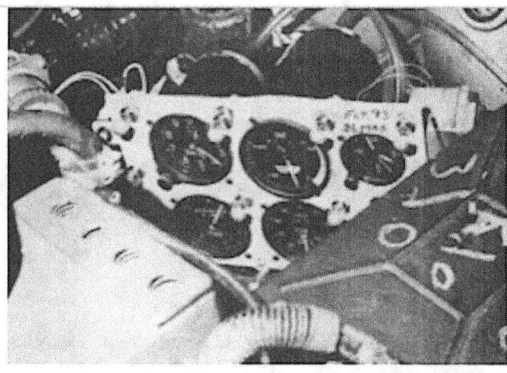

ADF Radio & IFF Transponder   Photo

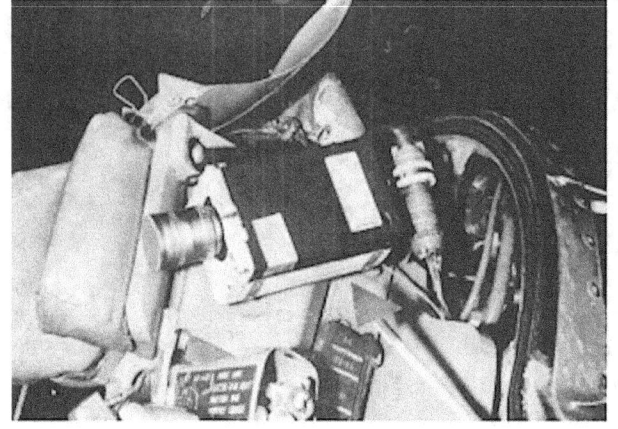

Recording Camera Left Side   Recording Camera

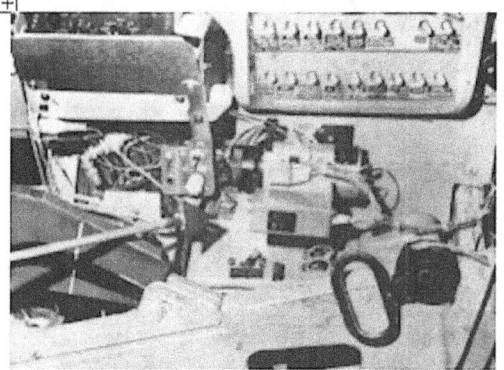

Photo Recorder   Abrams R-9A

Century Model 409 12-Channel   Oscillograph

Chapter 4 - Special Equipment Exploitation

For supersonic flight, the MiG-21 used a three position, translating, double angle inlet cone. The cone extended from the retracted position to an intermediate position at 1.5 IMN (Indicated Mach Number) and the full forward position at 1.9 IMN. The pilot manually positioned the usually automatic cone from the retracted position to the 1.5 or the 1.9 IMN position.

Even with comparable technology levels, the applications varied appreciably. The Soviet approach emphasized doing no more than necessary.

For example, in the low flow velocity corners of the intake ducts, the Russians did not countersink the rivets, leaving them protruding as much as one-eighth of an inch above the surfaces.

This, however, caused a negligible effect on the flow in the duct and made significant savings in construction effort to the Soviets.

Stowing the drag parachute cable channeled it through a shallow groove on the exterior surface of the aircraft along the ventral fin showed the simplicity and eased maintenance. Simple harp-shaped spring clips with thin safety wires across their tops held it in place.

The Soviet design concept 'wrapped' the smallest aircraft possible around the available power plant to assure maximum speed, altitude, and acceleration performance.

The Soviets carried this philosophy to the point where the fuselage bulges in various places provided clearance for equipment and accessories, rather than increase the fuselage diameter or cross-section area. The performance evaluation of the MiG-21 attested to the success of this approach.

The MiG-21 power plant, a Type 37P twin spool turbojet came equipped with a variable thrust afterburner. The engine, 181 inches long with a diameter of 35.7 inches, developed approximately 8,450 pounds of thrust at military power and about 12,650 pounds of thrust at maximum afterburner power.

The MiG-21 flight control system consisted of a manual rudder, hydraulically boosted manual ailerons and an irreversible horizontal stabilizer.

The main hydraulic system operated at 1200–1400 psi and powered the aileron boost and horizontal stabilizer.

The plane did not provide a backup manual control to the horizontal stabilizer, so to prevent over controlling at high-speeds. An automatic control altered the gear ratio from the control level to the stabilizer to decrease the range of deflection required of the stabilizer. However, it provided no stability augmentation, aileron trim, or rudder trim.

The MiG-21's design located its hydraulically actuated speed brakes on the underside of the fuselage, two forward and one aft, using a conventional tricycle-type landing gear with selectable two or three wheels, air-operated brakes. A lever on the control stick activated braking and a rudder that controlled the desired wheel.

The aircraft came equipped with a three-wheel, pneumatic, braking system. The added braking provided by the nose wheel brake making it very effective and further indicating the Soviet desire to operate their tactical fighters from relatively short runways.

A unique construction of the fuselage focused on construction expediency as well as weight savings. Rather than building up with frames and an inner and outer skin or honeycomb material, a single piece of metal composed the forward section.

US standards for tactical aircraft rated the range and payload capabilities of the MiG-21F-13 very low. However, the Soviets did achieve their apparent goal of developing a rugged, simple, highly reliable, and easily maintained fighter with the exceptional climb. Its altitude acceleration and maneuverability capabilities surpassed any other aircraft operational in early 1960. These characteristics remained excellent by tactical fighter standards if range and payload, not a prime concern.

The non-self-sealing bladder type fuselage cells with wing fuel stored in a "wet wing" provided a fuel system capacity of approximately 4,500 pounds of JP-5.

The aircraft used a simple, single button, self-contained, electrical battery starting system, and an air start system incorporated an autonomous oxygen supply designed for restarts up to 39,000 feet. Sufficient oxygen was available for four to five air starts of 30 seconds duration each. A special tank provided aviation fuel during the starting cycle.

The MiG Cockpit

The MiG-21 cockpit was armor plated, including armor plating installed behind the pilot's seat, forward of the instrument panel and aft of the forward windscreen.

Its sight system used an ASP-5ND lead computing gunsight. High Fix radar supplied range information in the cannon or rocket mode of operation to the gyro pipper. In the event of radar failure, fixed range inputs became available from 650 to 6600 feet.

For fixed armament, the MiG-21 had one NR-30, 30-mm cannon faired into the fuselage under the right-hand side of the pilot's cockpit. The cannon had a linear action with a mechanical feed chute, which roughly followed the contour of the aircraft's outer fuselage skin between the skin and the internal fuel tank. The cannon fired 850 rounds per minute with a muzzle velocity of approximately 2.500 feet per second.

The plane provided external stores on two wing-mounted stations and one centerline station. Each removable wing station carried one ATOLL missile, one bomb up to 1,100 pounds, or one l6-shot FFAR (Folding Fin Aircraft Rocket) pod.

Aircraft Modifications for Tactical Evaluation-Data:

The exploitation process modified the MiG-21 only where required for valid data acquisition and to conduct the technical and tactical evaluations.

The exploitation team installed UHF communications and removed the standard VHF equipment while adding a UHF blade antenna,

The exploitation team fabricated two wooden wing pylons to achieve representative combat and attached a LAU-7A missile launcher to each. The exploitation team then attached one AIM-9B training missile to each launcher.

The exploitation team placed a voice tape recorder on the right rear cockpit console and a communications lead that connected to the pilot's headset and microphone for providing necessary inputs,

Other modifications made for quantitative flight test data included a ten channel oscillograph, two over-the-shoulder cameras, a photo panel, X-band beacon, and a stopwatch. They used standard ITS instruments, airspeed indicator, calibrated Machmeter, and altimeter.

The ATOLL missile did not come with the plane, so for missile configuration, the exploitation team installed non-firing AIM-9Bs using an AIM-9 rail with laminated plywood/fiberglass and steel fittings. The substitution caused no performance change except slight improvement at low speed.

On-site Modifications

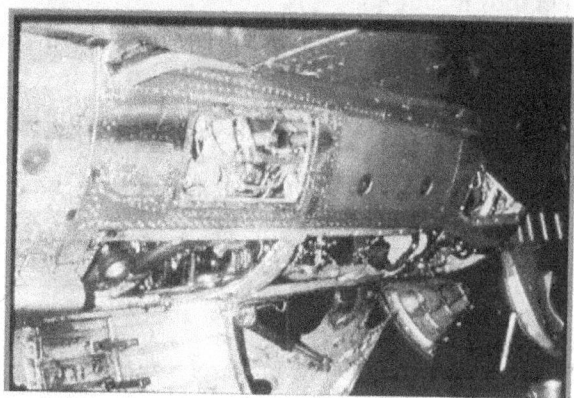

Radar cross-section Evaluation of the MiG-21

The radar ranges evolved through 2 phases. From 1959 until 1967 the ranges solely served to develop and measuring/verifying the A-12 Radar cross-section. With the deployment of the A-12, EG&G transferred many of its radar range personnel to NTS, leaving a few for a year to finish the closure.

This changed in the spring of 1968 with the arrival of the Project HAVE DOUGHNUT MIG-21 for exploitation. The aircraft arrived scheduled to fly over Tel Aviv in September that year to celebrate the win in the 1967 war in Israel. Because of it having such a short time frame, exploitation started the day it arrived.

The EG&G special projects team used six basic radar subsystems for cross-section measurements of the MiG tactical phase of exploitation.

An M-33, mobile, X-band fire-control radar, later replaced by the CIA's X-band Nike Hercules radar, tracked the target to generate range information and target bearings. A bull gear servo network manufactured in-house allowed the other radar systems to slave to Nike's target bearing data. The S-band radar with the 60 feet dish DYCOMS (Dynamic Coherent Measurement System equipment gathered the reflectivity data.)

The special projects engineers and technicians operated and monitored the individual radar performances from a master control facility using only four of the radar systems in the measurement program. The VHF, C-band, G-Systems detected radar return hot spots, while F-Systems made radar cross-section measurements. The RCS program followed the procedures established during the previous project OXCART's A-12 (Blackbird) articles in flight at the threat frequencies.

In preparation for this day, both Wayne Pendleton and TD Barnes spent time at Wright-Patterson

AFB. Pendleton briefed the Air Force Foreign Technology Division personnel on the range capabilities, and Barnes spent a year in the Flight Dynamics Laboratory to participate in developing a "black box" to evaluate Soviet MiG aircraft.

With most of the EG&G range personnel involved in Project OXCART already transferred to the Nevada Atomic Test Site, a much larger EG&G radar range contingent moved in for the MiG exploitation project. Besides having the MiG-21 on a short-term loan, the US Navy and Air Force desperately needed answers to stop the losses of flight crews to the MiGs in Vietnam.

Through the CIA EG&G special projects past 18 months of RCS evaluations of the CIA A-12 stealth plane, the team realized the target signature information as fundamental in any assessment. The radar signature identified the vulnerability of a weapons system to detection and tracking. Consequently, the Air Force element of the team sought both static and dynamic measurements of the MiG-21.

The special projects team accomplished this by using the Groom Lake ground-based VHF band, S-band, and C-band radar systems with the Nike X-band radar providing the primary tracking role.

During the dynamic RCS evaluation, the MiG-21 flew a flight pattern at an altitude of 30,000 feet and a speed of Mach.86.

Lacking telemetry, the special projects team could not determine the aircraft's attitude during turns. Consequently, the aircraft maintained a wings-level attitude during the straight portion of the flight.

The special projects measurement facility made measurements at 170, 2,900, and 5,050 MHz using six basic radar subsystems. The specially designed Nike bull gear servo assembly controlled their antennas with monitoring from the EG&G special projects control rooms. There existed the main control room and an auxiliary for special uses.

Four of the radar systems participated in the measurement program. The VHF, S, and C-Band equipment gathered the reflectivity data while the X-Band system generated range information and target bearings.

The other radars slaved to the reference servo network controlled by the target bearing data from the Nike X-Band radar. The special projects team digitized and recorded on tape the range-gated video data received by each radar system for off-line computer processing. For the MiG-21, they did not use the VHF telemetry system that normally provided target pitch, heading, and roll data.

**RCS broadside aspect
starboard turn**

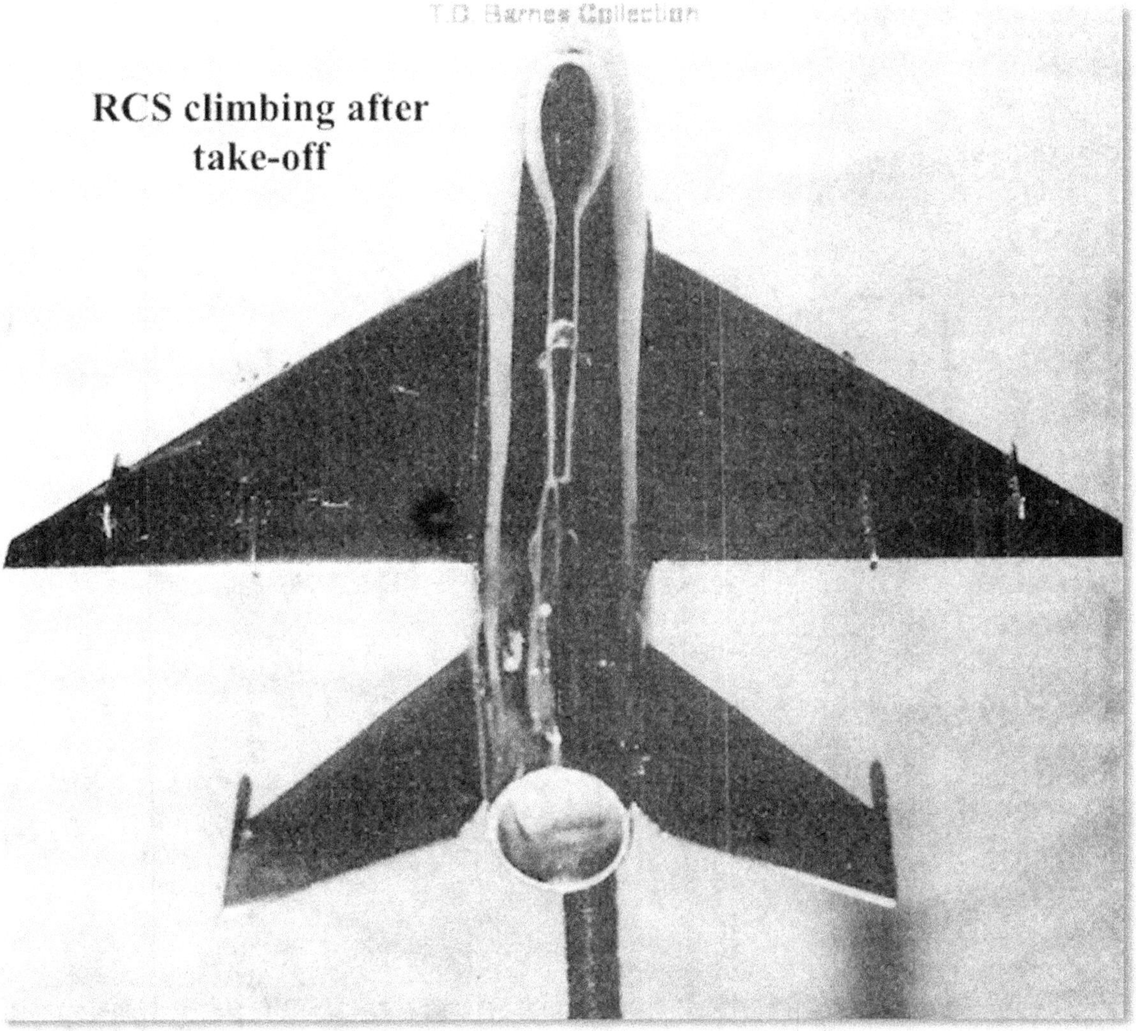
T.D. Barnes Collection
RCS climbing after take-off

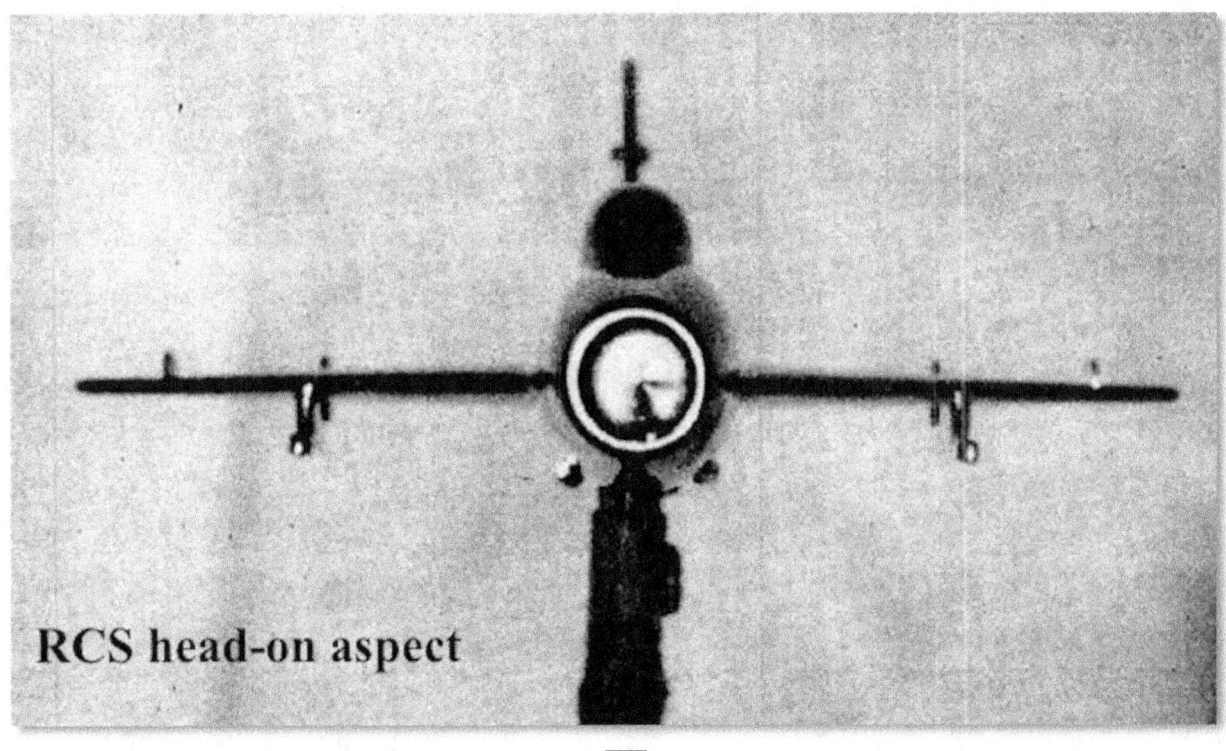

MiG-21 Fishbed on the Groom Lake Pylon

RCS head-on aspect

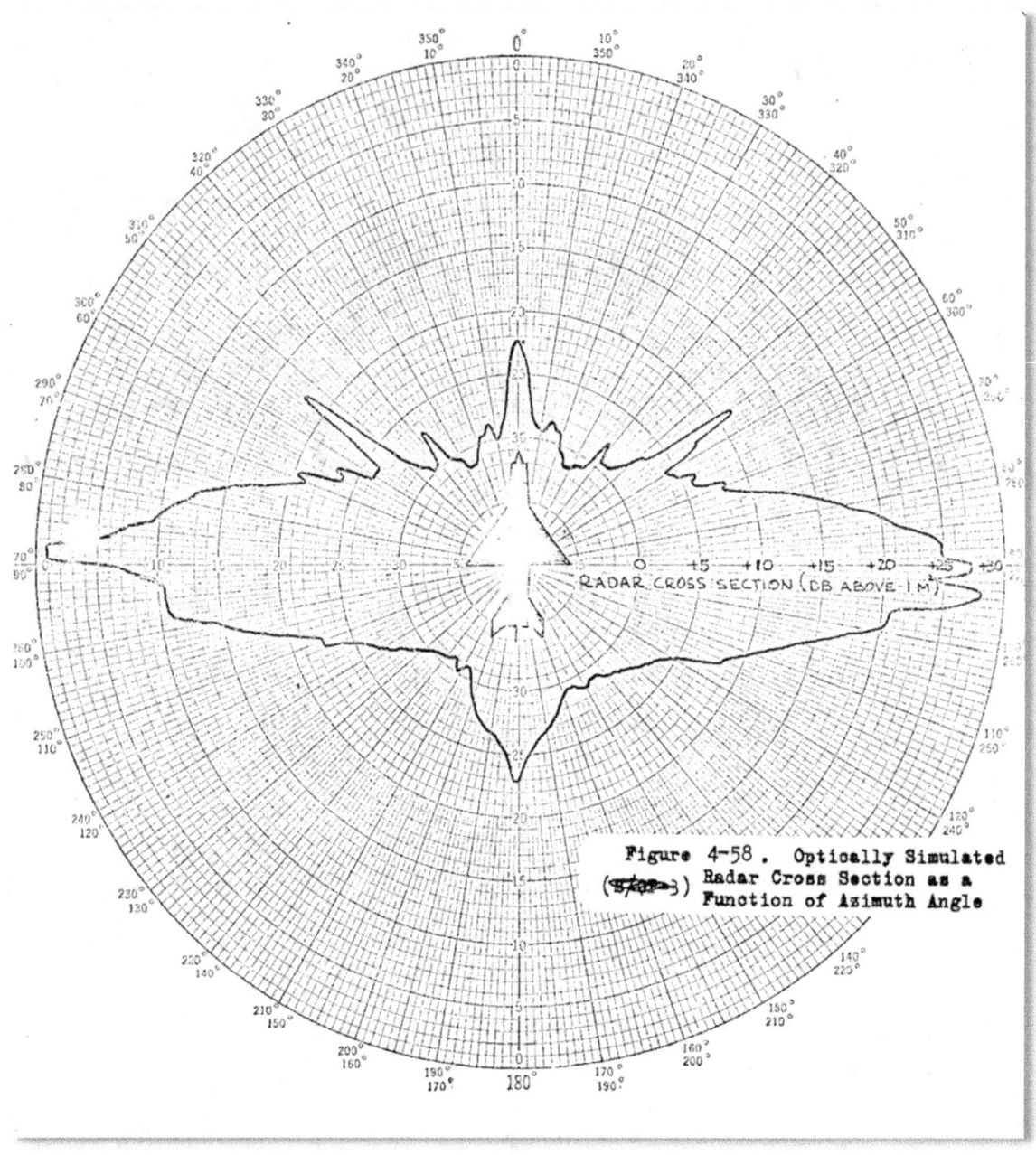

Figure 4-58. Optically Simulated Radar Cross Section as a Function of Azimuth Angle

Typical RCS scan of a plane on the pylon at Groom Lake by the special projects team.

Ground visibility tests

The Pilot Familiarization Flight

On 8 February 1968, excitement ran high throughout the exploitation team at Groom Lake. Today, they, representing America, prepared to fly the mystery plane, the crown jewel of USSR aviation.

The flight-testing of the MiG-21 began with the special projects exploitation team performing preflight on their radar systems. In the mission control room, Leroy Hardy and others situated the Air Force and Navy personnel at the various control consoles. Elsewhere on the base, the Air Force and Navy participants towed the MiG-21 to its take-off position at the end of the runway.

The tolling of the MiG exploitation vehicles to and from the active runway became standard procedure. It reduced the possibility of a ground mishap and preventing unnecessary brake and tire wear associated with taxiing. The entire exploitation effort gave primary consideration and awareness to the lack of available spares.

The exploitation team flew the MiG-21 in a natural finish with the number '90865' on the tail, and US national insignia on both sides of the nose, and none of the wings.

From their special projects position on the edge of the dry lakebed, the exploitation team participants first heard the start of the engine of the chase plane as it taxied to the runway. The sounds of the participating plane starting up at the end of the runway followed.

The screaming jet sound of the chase plane's engine and then the run-up of the MiG-21 engine alerted the entire flight test facility of the Soviet MiG beginning its flight.

From that point, the small, secret flight test facility enjoyed viewing an aerial show. The exploitation team put the MiG-21 through a joint TAC and USN choreographed first flight of the MiG-21 Fishbed flying over Nevada to familiarize the pilot with the plane. Little did any of them know that Area 51 became a permanent flight test facility with that first scream of a Soviet MiG engine racing down its runway.

MiG-21, F-8E, F-4E

The MiG-21 in Hangar Five at Groom Lake. Note the US Air Force decals. The Air Force designated the plane YF-110B.

The familiarization flight lasted thirty minutes with the MiG-21 and F4D conducting two acceleration and two deceleration runs.

Evaluation started on the ground evaluation with start, taxi, and takeoff, where TAC found the MiG-21 engine response poor during taxiing and engine checks.

The pilot found the wheel brakes only fair and steering difficult with differential braking. The wheel brakes failed to hold during the run-up at full power.

The MiG-21 demonstrated good acceleration on takeoff and stabilator effectiveness on rotation. However, the landing gear failed to retract until the third recycle attempt with slight trim change during gear retraction.

Maj Gerald D. Larson, TAC, observing Fred Cuthill preparing for a test flight at Area 51

The MiG-21 experienced a slight sink during flap retraction.

The pilot found the speed brakes performed poorly and ineffective, and aileron control was very sensitive at low speed. The pilot experienced pronounced adverse yaw during low-speed maneuvering.

The afterburner refused to ignite until engine speed reached 100%, and the plane encountered airframe buffeting about 550 KIAS, which stopped with power reduction and deceleration.

During the preflight familiarization, the Navy found the pilot strap-in cumbersome, requiring the assistance of a plane captain.

The simple prestart cockpit checks still required concentration due to the cluttered switch panels and similarity of switches. The air operated canopy required approximately 10 seconds to close after actuation, locking and seal pressurizing manually.

The canopy required closing before moving the aircraft. The Navy also reported the pilot's visibility aft severely limited by the ejection seat headrest, windscreen, and armored glass combination.

A typical mission started with Barnes locking the Nike X-band radar onto the MiG-21. The rest of the radar systems tied to this radar by the bull gear designed strictly for RCS evaluations. Each of the other participating radar systems locked to the azimuth and elevation of the Nike radar.

In a planned exercise, the MiG-21, along with an Air Force McDonnell Douglas F-4D Phantom II and a Navy Vought-produced F-8E Crusader, climbed to 10,000 feet. At the word, go, they conducted an acceleration comparison.

The "black box" data collection system designed and placed on board the MiG-21 for this purpose evaluated the afterburner, recorded the engine response, aircraft maneuvering qualities, slow speed handling characteristics, avionics, and sight system analysis.

Inside the special projects building, those monitoring the control consoles, radar, and data systems proceeded with a maneuvering flight of the MiG-21. It entered a nose lightening occurred at about 5.5 G accompanied by high airspeed bleed off.

The pilot reported it as having a medium to heavy stick forces as it approached the stall where the plane experienced mild buffeting and wing rock. The pilot affected recovery at 140 KIAS as the left wing dropped. While accelerating from low speed, he reported the intake suck in doors closing with a noticeable bang. Below 200 KIAS in the traffic pattern, the aircraft felt sensitive to the controls.

From the TAC flight control console, a TAC officer directed the MiG-21 to ascend to 10,000 feet for an acceleration check with the F-4-D chase aircraft.

During the run, the F-4 maintained a superior performance throughout (300 to 400 KCAS) in military power. Following that run, the planes conducted an afterburner acceleration check from 300 to 550 KCAS. The F-4E maintained a wing formation position with excess power available to separate from the MiG-21. At 550 KCAS, the MiG-21 terminated the acceleration because of severe buffeting.

A level flight deceleration with speed brakes followed that found the MiG-21 and F-4 equal in speed brake deceleration. However, at idle power, the F-4 decelerated more rapidly. Slow speed maneuvering required proper piloting technique because of wing roll off and adverse yaw characteristics of the MiG-21. The MiG pilot reported poor visibility through the forward windscreen with targets acquired at a 3 to 5 NM range. The canopy also restricted rearward visibility to where the pilot saw only about one foot of each wing tip.

The MiG-21 pilot reported that selecting afterburner by moving the throttle into afterburner range from any position less than full military RPM delaying ignition until the engine accelerated to 100%. The F-4 chase pilot reported no apparent afterburner puff during Air Base operation at 10,000 feet and below and no apparent engine smoke at any time. Ground photos of the planes during maneuvers confirmed what the chase pilot reported.

The three planes grouped at 12,000 feet where the MiG-21 made a level CRT (Combat Rated Thrust) acceleration from 300 KIAS to 450 KIAS. The F-4D, flying alongside, passed the MiG-21, however, the F-8E, in the trail, failed to match the MiG-21's acceleration time of 50 seconds and dropped behind.

The planes conducted a level CRT acceleration at 12,000 feet from 300 KIAS to 550 KIAS with afterburners selected on a signal from the special projects' control console manned by TAC and Navy mission controllers. The MiG-21 afterburner ignition delayed approximately 9 seconds with no afterburner puff observed.

The F-4, flying alongside passed the MiG-21, using half-modulated afterburner to maintain position. The F-8E MRT (military rated thrust) acceleration rated both equal and superior to the MiG-21. The MiG-21 deselected afterburner at 550 KIAS due to engine surging and airframe buffet with published Vmax (maximum speed) at this altitude as 595 KIAS.

The plane's speed breaks decelerated like F-4D and MiG-21 deceleration characteristics. The F-8E deceleration characteristics functioned superior to the MiG-21.

The MiG-21 initiated a 180-degree turn at 550 KIAS and 5.5 g with adequate horizontal stabilizer available throughout the turn. It experienced left wing dip and longitudinal control lightening at 5.2 g.

The Navy F-8E held a clear AIM-9D tone on the MiG-21 at 1.5 miles out to a 60-degree TC. The planes performed low speed turns to buffet at 250 KIAS (1.8.g left-2.0 g right) where the MiG-21 experienced light buffet at 2.2 g.

The planes next performed straight and level, clean stalls at 78 percent RPM where the MiG-21 encountered light buffet at 185 KIAS, wing rock at 175 KIAS and left wing roll off at 150 KIAS. It lost approximately 500 to 800 feet in stall recovery due partially to slow engine acceleration. It recovered in a slightly unbalanced flight and rolled off in the direction of adverse yaw at 175 KIAS.

The pilot of the MiG-21 felt comfortable at 300 to 500 KIAS. Below 200 KIAS, he reported the aircraft feeling squirrelly. During the later debrief, the pilot stated his feeling that only a well-experienced pilot should attempt scissors, reversals, etc., in the MiG-21 to maintain completely coordinated control in this area. He recommended pilots avoiding unstable flight and subsequent departure.

The planes made an uneventful return to the flight test facility with a low approach followed by a final landing. The MiG-21 pilot used his speed brakes to maintain high engine RPM during the approach, extending his landing gear and flaps to 250 KIAS. In the landing configuration, longitudinal control lightening occurred at 140 KIAS followed by wing roll-off, flying the flight test facility leg at 200 KIAS with 170 KIAS on final with a normal touchdown.

The drag chute actuated after touchdown at 140 KIAS and deployed at 110 KIAS. The pilot selected two-wheel braking before turning off the runway. Engine shutdown was normal.

During the debrief, both the Air Force and the Navy found the MiG-21 extremely small and difficult to see. Unlike American planes, it left no apparent smoke trail at any power setting.

On 11 February, the exploitation teams conducted a second familiarization flight that lasted 35 minutes with the MiG-21 and F-4 conducting two acceleration runs while evaluating handling characteristics, avionics equipment, and climb comparison.

Chapter 5 - MiG-21 Tactical Evaluation

Moving into the tactical phase of exploitation, the National Air and Space Intelligence Center, along with the EG&G Special Project team, conducted choreographed USAF and USN combat flights from the special projects mission consoles. There, they evaluated the effectiveness of existing tactical maneuvers in an air-to-air environment. They optimized existing tactics and developed new tactics while evaluating the design, performance, and operational characteristics of the MiG-21.

For offensive and defensive evaluation, TAC evaluation aircraft flew 35 of the 102 total sorties on the MiG-21. The MiG-21 flew against the F-4C/D/E, F-105D/F, F-111A, F-100D, F-104D, F-5A. The RF-101, RF-4C, and B-66 flew defensive evaluations only.

The special projects personnel supported the exploitation team in acquiring maximum documentation during the data acquisition phase. Also, the safety chase aircraft, participating aircraft, and ground monitor used gun camera and external pod-mounted 16 mm Canon Scopic motion picture cameras when possible on selected flights.

The exploitation team installed over-the-shoulder cameras in the MiG-21 to record cockpit conditions during each flight cockpit. The MiG-21 used voice tape recorders, safety chase, and participating aircrews to document each tactical air situation and development. Ground monitors recorded the UHF communication during comparative flight evaluations. The exploitation team recorded and summarized briefings and debriefings daily. All participating aircrews completed flight data cards during or immediately after each flight with significant events and pilot qualitative comments noted.

MiG-21 Fishbed-Exploitation Maneuvering at Area 51

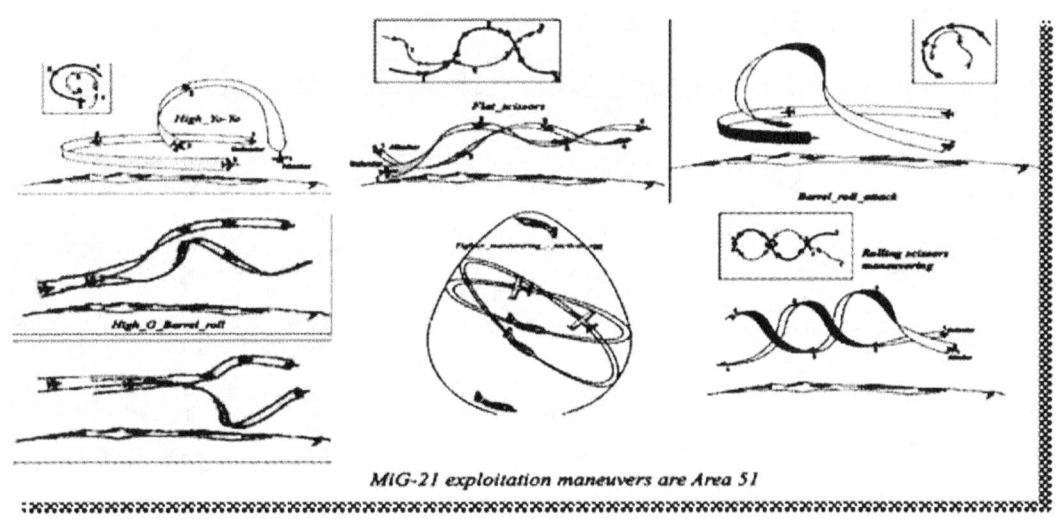

MiG-21 exploitation maneuvers are Area 51

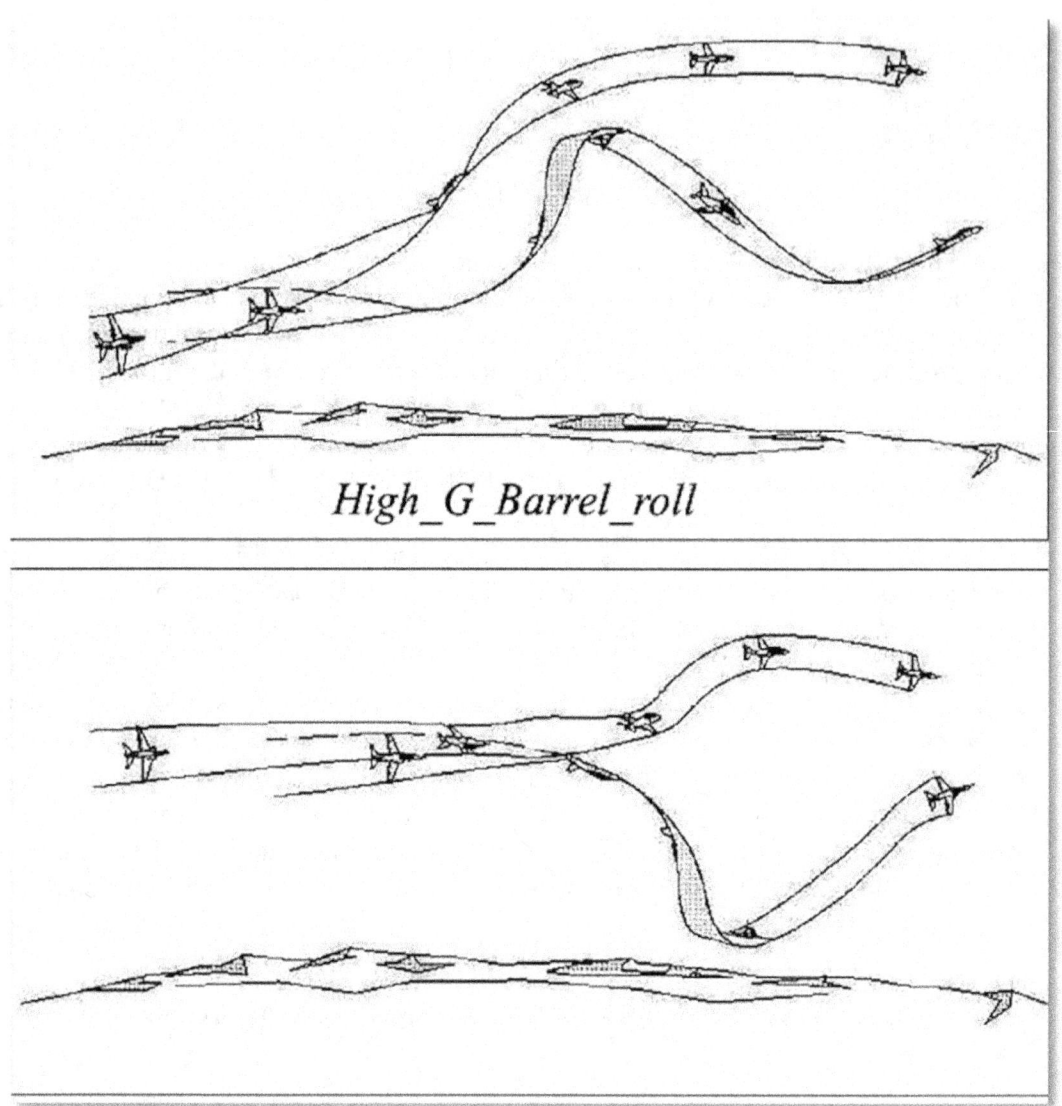

High_G_Barrel_roll

F-5N

F-101

B-66

F-100D

F-104

TAC F-4C/D/E

F-105D

F-111A

Navy F-4 Phantom II

Navy F-4J EI

A-6E Intruder

A-7E Consair II

Navy F-8E Crusader

Air Defense Command F-106

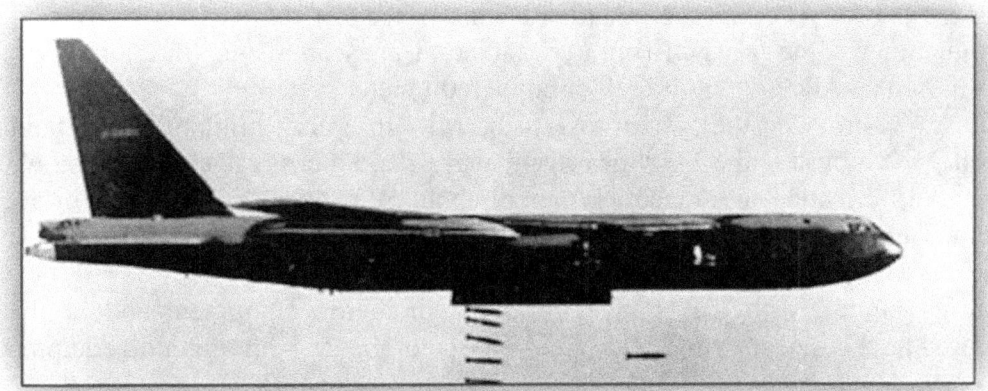

Strategic Air Command B-52

HAVE DOUGHNUT Tactical

A-4F Skyhawk

Joint Air Force and Navy Technical Evaluation Flights on the MiG-21

During evaluation of the MiG-21 infrared radiation measurements, the joint Air Force-Navy exploitation team used an APGC ATR-1/T-39 airborne radiometer system.

China Lake NWC provided the system to test the MiG in cruise, military, and afterburner modes. Wright-Patterson AFB furnished the T-39 measuring aircraft and a chase plane equipped with a standard IFF transponder took formation in a racetrack pattern about 50 nautical miles long. For acquisition, the MiG-21 positioned about one mile in front of and slightly above the radiant-intensity measurement aircraft with the chase aircraft on the wing of the MiG.

The measuring aircraft with the Navy's infrared radiometric equipment and IF missile seekers established infrared tracking of the MiG with the ATR-1 radiometer while ranging equipment locked on the IFF.

The pilot of the MiG provided pertinent performance data by UHF communication on the MiG's pertinent performance data through various power settings and maneuvers.

Among the many radiation characteristics of the Fishbed-E aircraft, the most interesting data revealed no significant increase in values during altitude increase or at extended slant ranges.

The exploitation team successfully conducted a ground-based IR measurement program against the MiG-21 Fishbed-E both on the ground and in the air. The team used an infrared television, and two infrared radiometers, a CVR (circular variable filter) spectrometer, and a movie camera.

They collected data during engine run-ups with the aircraft sitting at the end of the runway before takeoff. They did the same during takeoffs and landings, and during flyovers, recovering useful data for infrared search, warning receivers, and similar equipment.

The IRTV sensor successfully detected the Fishbed in tail aspect and without afterburner at 40 KM. They found the exhaust plume of the Fishbed-E significantly smaller than the exhaust plumes of F-4 and F-8 aircraft in the same wavelength region.

APGC ATR-1/T-39 airborne radiometer system

The Wright-Patterson Air Force Base T-39 twin-jet aircraft served as an airborne platform to provide the infrared color-wheel radiometers and for the missile-tracking display instrument provided by NWC. The T-39 was a very reliable platform able to stay aloft for three hours and fly to an altitude of 45,000 feet at a maximum speed of Mach .8.

The exploitation team removed the T-39 starboard passenger seats to prepare for a test flight and replaced the original starboard escape hatch by either of two modified by NWC. One accommodated a turret designed for use with the radiometers, and the other accommodated a fiberglass turret used with the missile-tracking-display units.

They fitted each turret with infrared-transmitting windows mounted on gimbals to permit wide-angle manual tracking. Snap-pins ensured quick installation and proper alignment of the instruments in the turret. The exploitation team placed the radiometer power supply and a signal-monitoring oscilloscope on the deck and secured it to the bulkhead with a metal snap band. A nearby panel provided 400-cycle aircraft power and transferred warm air to the turret windows from aircraft heating ducts through a flexible rubber tube to prevent the formation of frost.

The exploitation of the IR detection of the Fishbed-E found it more susceptible than either the P-4 or P-8 aircraft. They found this especially so during a tail attack by IR air-to-air or ground-to-air missiles operating in the 2 to 4-micron region. The smaller intense plume of the Fishbed-E provided better guidance characteristics to the missile since it more closely resembled a point source. A missile guided performed better on the smaller Fishbed-E plane, making it much more susceptible to contact or influence fused TR missiles than the P-4 and P-8 aircraft.

Where the spatial filtering provided by the reticle in an IR missile guidance head specifically rejected extended area (i.e., clouds, etc.) targets, the Fishbed-E required a more intense flare than either a P-4 of P-8 aircraft to successfully flare decoy an IE missile. A missile launched against a Fishbed B saw a small area plume. This made it less apt spatially to reject it for a point source flare. A smaller miss distance to the missile ensured having the MiG-21 in lethal range when the fuse activation/warhead detonation occurred.

The AIM Guided Missiles

The AIM-7 missile was a supersonic air-to-air homing missile substituted for the ATOLL missile during HAVE DOUGHNUT. The launching aircraft supplied guidance by the transmitted CW signal. To lock-on and track the target using proportional navigation, the missile compared the CW signal received from the launching aircraft with a CW signal reflected from the target.

The supersonic A1M-9B/D air-to-air homing missiles employed passive infrared target radiation for guidance also substituted for the ATOLL missile. The missiles use the infrared energy emitted by the target to lock-on and track the target. Target detection provided an audible tone to the aircrew when achieving target detection.

Radar Homing and Warning Set AN/APR-25

The AN/APR-25 is a passive ECM (Electronic Countermeasures) set and consists of warning receiver, threat indicators, and an audio system. The radar homing and warning set indicated threat bearing about aircraft heading, relative signal strength, and the type of emitting activity. It discriminated threat signals from non-threat signals by radio frequency signal strength, pulse repetition frequency, and antenna scan rate. Threats displayed on the strobe display scopes and threat indicators in each cockpit.

The MiG-21 30 mm Cannon

The MiG-21 30 mm cannon was 100 percent reliable. The gun fired on five different sorties. On each sortie, they loaded and fired ten HEI (high explosive incendiary) rounds in a single burst in near one g flight at a ground target with no jams or failures to fire encountered. They found the muzzle flash readily visible at one mile in daylight.

The exploitation team demonstrated the lethality of the 30-mm cannon in a ground attack flight conducted against a standard.

The exploitation team demonstrated the lethality of the 30-mm cannon during a simulated ground attack flight. The target for the strafing attack was a standard; the US manufactured bulldozer. The weapon rendered the bulldozer inoperative and irreparable after hitting it with a two1 round of 30 mm HEI ammunition. The 20-mm cannon used by US tactical fighters failed to cause a comparable degree of damage.

MiG Targets at Area 51

▪ USAF Tactical	33
▪ USN Tactical	25
▪ USAF Performance, Stability & Control	26
▪ Air Defense Command	4
▪ Strategic Air Command	2
▪ Infra Red	9
▪ Radar Cross Section	1
▪ Photo	1
▪ Acceptance Flight	1
▪ Total Sorties	102

Weapon targets for MiGs at Area 51

The MiG-21 Armament and Fire Control System

The exploitation of the Fishbed-E Armament and Fire Control System found that it suffered from its forward visibility-degrading the overall effectiveness of the Fishbed-E weapons system. This required the pilot to retain visual contact with the target to employ the tracking index, fire control systems, and armament effectively. The sight combining glass, the bulletproof glass slab, and the forward windscreen frequently caused targets unobserved, or not visually spotted.

While evaluating switchology, during air-to-ground attack flight, the pilot had enough time to position switches as required. However, the pilots found the air-to-air switchology requirements initially to set up the systems more difficult. Setting up excessive pilot actions. It was possible, however, to convert from a missile attack to a gun attack with one switching movement.

The armament of the MiG was surprising. The HR-30, 30 mm cannon was limited to a capacity of 60 rounds with an estimated gun rate of 850 rounds/minute, which provided for a total firing of 4.2 seconds. Estimated muzzle velocity was about 2,560 feet/second. During the gunfire, pipper jitter was excessive, about 20 mils, and tracking correction during gunfire impossible. The participants saw the muzzle flash during daytime conditions from a range of 3 miles.

The exploitation team used a bulldozer as a target during the tactical phase of the exploitation. The bulldozer was in operational condition before this attack, although the exploitation team removed the blade. Estimated two rounds of HEI impacted the vehicle and rendered it irreparable. The evaluations encountered no cannon malfunctions during the cannon firing missions.

Manual ranging of the gunsight could not perform smoothly and precisely. System hysteresis-and friction made it virtually impossible to prevent over control of the sight reticle diameter size with the throttle twist grip, pipper jitter during cannon firing was exceeding 20 mils.

Gyro drift when tracking air targets was excessive. G loads greater than 2.5 caused the sight reticle to drift to a point near the bottom of the sight combining glass. At very high g loads, the sight reticle disappeared entirely.

Sight electrical paged function was sluggish and slow to respond. During air-to-air tracking, it was necessary to hold the electrical cage button (on the stick grip) until radar lock-on occurred. The electrical page button was poorly positioned and difficult to actuate when preparing to fire the gun.

The Fishbed's over-the-nose visibility restrictions limited the useful mil depression to 94 mils.

They found no large lead angles available during air-to-ground attacks with bombs, gun, or rockets, and it impossible to depress the gunsight in the gun mode of operation required for the ground attack at long slant ranges.

The exploitation did find some desirable characteristics incorporated in the MiG-21F-13 Fishbed C/E. Weapon System. This included ease of system operation, simple cockpit procedures, and minimum system monitoring enhance pilot performance during a tactical engagement.

The small frontal area provided a low probability of visual or radar detection of the MiG-21 in a head/tail-on aspect. The MiG-21 pilot could use the quality to his advantage for reduced detection during patrol or attack. After initial visual detection of the MiG-21, it was necessary to "padlock" or remain visually fixed on the aircraft to prevent losing contact with the US tactical aircraft, the F-104 as it compared in size.

The operational weight of the Fishbed, configured with 60 rounds of 30 mm ammunition and 2 ATOLL missiles is 16,250 pounds. Although not demonstrated, it was possible to operate the MiG-21 weapon system from soft runways, i.e., snow, dirt, sod, etc. The main gear tire consumption during 50 landings rated comparatively low and normally available. During this evaluation, one set of tires accumulated 53 landings.

The MiG-21's head up display provided the pilot with target radar lock-on, target in range, and over-g condition for a missile launch, and target breakaway (minimum range). Although lacking in sophistication, the presentation provided the MiG-21 pilot with required information in a manner, simple and effective.

TAC Inventory Versus MiG-21F Fishbed flight Test Overview
The joint TAC, USN, CIA EG&G special projects and other government agencies' tactical exploitation phase flew 40 of the available 52 days.

They canceled eight days due to weather and four days due to maintenance. Of the 134 sorties scheduled, the tactical phase flew 102 sorties, canceling 21 sorties due to weather and 11 due to maintenance.

The flights consisted of one acceptance flight, one photo, 1 RCS, 9 IR, 2 SAC, 4 ADC, 26 performances, stability, & control, 33 TAC Tactical and 25 USN tactical. The Air Force and Navy flew exploitation and evaluation flights both solely and jointly.

Various customers in the exploitation desired and conducted their tests using their aircraft and seeking solutions that differed from the other participants. Therefore, where there might be some duplication, it is because this book covers each participant's interest, tactical evaluation, and conclusions.

During the exploitation of the MiG-21, the USAF and USN evaluated the aircraft performance, aircraft stability and control, armament, and cockpit environment again some planes. These included the F-4C/D/E, F-105D/F, F-111A, F-100D, F-104D, F-5N, RF-101 (defensive only), RF-4C, and B-66 (defensive only).

Sortie Breakdown

Emphasis on What the Air Force and Navy Learned from Matching the F-4 Against the MiG-21
From a combat spread of 1 mile, the F-4 section meeting the MiG-21 head-on immediately selected afterburner, accelerated, and separated, affecting a VID maneuver, or commenced loose deuce maneuvering.

The maneuver forced the MiG-21 to pick one F-4. At that point, the engaged F-4 completed a head-on pass followed by an oblique loop as described in the one-on-one tactics that freed the second F-4 to maneuver and immediately press for the offensive.

Head-on passes by the F-4 kept the MiG-21 engaged while the free F-4 maneuvered into a missile launch position. It was essential for the "free" F-4 to maneuver rapidly in a different plane and strive for a rear quarter attack. If the MiG-21 switched to the free F-4 during the engagement, the F-4s also switched positions, making the previously engaged F-4 the free F-4.

When both F-4s reached an astern position on the MiG-21, the basic tactics previously described applied to the attacking F-4. While one F-4 committed to attacking the other positioned himself out of the plane of the maneuver, preferably in a high cover position, and be ready to conduct a slashing attack. Since the MiG-21 had a high rate of turn and small turn radius, an F-4 high yo-yo easily resulted in a head-on pass coming down from the apex of the yo-yo.

The exploitation team was proficient in ACM and familiar with the maneuvers described in the F-4 Tactical Manual that describes defensive maneuvering by the attacked F-4 in one-on-one tactics. Early separation in the vertical by the free F-4 provided mutual support was necessary to gain a missile launch position and sandwich the attacker. If the MiG-21 switched and attacked the higher F-4, the high F-4 broke down into the attack. He informed his exploitation teammate of the switch and directed the exploitation teammate to ease turn before executing on oblique loop.

Passing through the vertical, the pilot sighted his exploitation teammate passing underneath on a near reciprocal heading, with the MiG-21 pursuing or remaining high and switching his attack. If the MiG-21 continued pursuing the low F-4, the low F-4 continued separating while the high F-4 completed the loop behind the MiG-21. If the MiG-21 remained high, the high F-4 called the switch and directed the free F-4 to execute an oblique loop. The engaged F-4 generated a large overshoot, then dove for sufficient separation to allow a reversal for a head-on pass, while the free F-4 maneuvered for the kill.

The same basic tactics applied if encountering multiple bogies. Section integrity, lookout doctrine, and mutual support remained mandatory. The more complicated the tactical situation, the tighter the section maneuvering became. The split plane, vertical maneuvering continued once the attacked bogey aggressively maneuvered defensively or attacks the section.

To eliminate tracking problems through MBC after initial detection in pulse Doppler, F-4J aircrews initiated rapid relock to the pulse mode, initiating rapid relock early, contingent on the following factors:

- Delay rapid relock on a suspected MiG-21 target until the range is well inside the expected detection range of the MiG-21 by the pulse radar system.
- Differential altitude. Target look down angle eliminated before initiating rapid relock.
- Intercept geometry. Stopping target drift before a rapid relock attempt indicated that target contact was lost in main beam clutter unless a rapidly changing Vc, making it mandatory to make an immediate attempt to relock in pulse mode.

During evaluating target maneuverability, pulse relock eliminated the possible loss of radar illumination as a maneuvering target enters the main beam clutter notch. Ideally, the flight accomplished pulse acquisition and acknowledgment before 20 nautical miles during a VID maneuver.

The continual center of the split elevation strobe and the expanded velocity display was mandatory before rapid relock to Pulse. Periodic use of the pulse mode made during BARCAP/TAR-CAP operations to offset the difficulty in detecting low Vc targets, i.e., Vc lower than F-4J TAS.

The extensive turbine modulation of the MiG-21 was a possible aid to pulse Doppler detection of low Vc MiG-21 targets. Aircrews remained aware of the erroneous Vc information presented and used it to assist in rapid relock.

Whenever possible, avoid assignment of co-channel and adjacent channel APG-59 radars to the same section or CAP station. When unavoidable, exclusive use of the pulse Doppler mode was not reducing the effect of mutual interference.

F-4s equipped with pulse-only radars stayed aware of the short-range contacts probable during overland operations at current BARCAP/TARCAP altitudes. In an area of known MiG activity, visual search, and random heading changes aimed at thwarting a GCI-controlled MiG-21 intercept took precedence over radar search.

Due to the pilot field of view restrictions in the F-4 and the small size of the MiG-21, it was essential that RIO's use "padlock" lookout technique while engaged in ACM. The pilot did not attempt radar acquisition until the MiG-21 was within approximately 45 degrees of the nose and confirmed visual contact to eliminate unnecessary position calls that might block out UHF transmissions from the accompanying F-4.

Rios remained thoroughly familiar with all phases of ACM to assess an enemy maneuver and provide proper directive commentary if the pilot lost visual contact for a protracted period. During actuation of PLM, the RIO switched his attention to the radar scope and assisted the pilot in determining the proper direction of range gate sweep, based on the relative positions of the target and altitude line. After lock-on, the RIO verified valid target track by noting correct Vc, illumination of range track light and AIM dot, and elevation strobe deflection. He notified the pilot and called the range to the target. On obtaining a false Lock-on, he broke the lock and told the pilot to reacquire. Timely and accurate information from the RIO reduced the possibility of launching an AIM-7 out of the envelope.

Due to rapidly changing target aspects, the pilots used the pulse Doppler mode only as a secondary mode of operation during ACM. If conditions warrant, they will employ automatic acquisition. However, they anticipated spurious automatic lock-on on other aircraft in the area or beyond visual range.

In an area with some friendly I-band emitters present, visual and radar search took precedence over APR-25 indications for evidence of MiG-21 activity. Under no circumstances did they ignore an APR-25 I-band track indication of an unknown source. Conversely, attention to APR-25 search did not detract from other means of detecting the presence of MiG-21, i.e., visual and radar search.

A MiG-21 was most difficult to acquire visually more than two nautical miles. For this reason, a rigid section lookout doctrine was mandatory if the section expected to operate successfully in a hostile area. Section loose deuce maneuvering, separated in the vertical plane, was most effective formation for converting a defensive situation to the offensive.

Attack aircraft employing Tactical Manual defensive maneuvers against the MiG-21 proved effective. The exploitation team validated all maneuvers in one-on-one engagements and stressed lookout doctrine. Maintaining a tactical section provided mutual support and to enhance lookout doctrine.

USN MiG-21 Findings

The Navy concluded the MiG-21 was extremely difficult to detect and keep track of in the ACM environment visually. The MiG-21 had a definite tactical advantage due to its small size and was a highly maneuverable aircraft capable of high g, low-speed flight with a Mach 2 capability at high-altitude.

- The MiG-21 size and shape looked very like an A-4. The A-4, A-6, A-7 possessed sufficient maneuverability in an initial break turn to thwart a MiG-21 attack. However, resulting in energy loss allowed MiG-21 to reattack with comparative ease if the A-4/6/7 stayed in the fight at slow speed. The MiG-21 pilot could choose to engage or disengage the A-4, A-6, A-7 aircraft at will. The A-4 and A-6 could reverse and obtain a quick snapshot if the MiG-21 overshot close in. The A-7A lacked the necessary thrust to complete this high nose maneuver.
- A clean MiG-21 met with heavy airframe buffet at.96 IMN below 16,000 feet. However, high stick forces limited its maneuvering flight characteristics at speeds above 510 KIAS below 16,000 feet.
- The MiG-21F's turning ability was impressive due to low wing loading and high thrust to weight ratio. Flown to maximum performance, it out turned an F-4 and F-8 series aircraft in a close-in turning engagement.
- The MiG-21 zoom performance up to 25,000 feet was inferior to the F-4 fighter configured without a centerline tank.
- The MiG-21's zoom performance was comparable to an F-4 fighter configured with a centerline tank.
- The MiG-21 zoom performance was comparable to the F-8 below 25,000 feet.
- The MiG-21 on station ACM time was comparable to the F-4/F-8 series aircraft with a similar percentage of total fuel on board.
- The MiG-21 gun sight system tested had limitations in the ACM environment.
- The MiG-21 airspeed bleed off in a high q turn was rapid below 400 KIAS.
- The MiG-21's total performance degraded only slightly with the installation of Sidewinder type missiles.
- The MiG-21's engine acceleration was slow from 85 percent to 100 percent.
- The MiG-21 leaves little or no smoke trail at military or afterburner power settings.
- The MiG-21 cockpit visibility suffered serious degrading through the forward windscreen,

below the canopy rails, and in a 50-degree cone aft
- The MiG-21 30 MM cannon effective and reliable.
- The MiG-21 an extremely vulnerable aircraft
- The MiG-21 easy to maintain.
- The MiG-21 sortie rate was high.
- The MiG-21 could recycle in 30 minutes.
- The F-4/F-8 series aircraft had a tactical disadvantage in the ACM environment because of their large size and prominent smoke trails.
- F-4/F-8 series aircraft capable of exceeding the MIT–21's q limit at low altitude.
- F-4/F-8 series aircraft had better CRT acceleration performance than the MiG-21 below 1.2 IMN at low and medium altitudes.

MiG-21 & F-8 in flight over Groom Lake in Nevada

Recommendations
The Commander Operational Test and Evaluation Force recommended that:
- Navy fighter aircraft engage a MiG 21 in section.
- Maintain section integrity throughout the engagement for mutual support.
- Always maintain a strict lookout, and the engaged aircraft use "padlock" lookout technique.
- In a threat area, weave and vary headings along the base course.
- Force all engagements to low altitudes at high-speeds.
- Determine the MiG-21 pilot's ability early in the engagement.
- A close-in, avoid slow speed engagement if the MiG-21 is flying at or near the maximum performance.
- Orient offensive maneuvering toward exploiting MiG-21weaknesses rather than rushing for a quick kill.
- Keep an attacking MiG-21 at high TC.
- Aircrews be aware of situations or conditions that warranted disengaging from a MiG-21 encounter.
- Employ only sound, proven tactics.
- Practice ACM under controlled conditions against small aircraft with low wing loading, e.g.,

A-4F, E-5.
- Aircrews be thoroughly familiar with and aware of their weapons system limitations.
- Exploit the limitations of the enemy.
- Whenever possible, Radar Intercept and air combat conduct training over land.
- Make AIM-7E attacks on a maneuvering MiG-21 in the pulse mode of the APG-59 radar.

The USN concluded the aircraft easy to fly with no dangerous characteristics. The design of the vehicle avoided complexity occurred whenever possible. Particularly noteworthy was the plane employing no stability augmentation.

The acceleration and thrust-limited turning performance of the aircraft, although less than predicted, was good throughout the flight envelope.

Basic aircraft stability, except lateral-directional damping, was good. The aircraft exhibited excellent lift-limited maneuvering characteristics for both the available load factor and handling qualities near the stall. The MiG-21 exhibited good roll rates and response through the flight envelope.

In turbulent conditions, the aircraft failed as an acceptable platform for weapons delivery or instrument flying because of weak lateral-directional damping combined with slow engine response.

At low altitude in the transonic region (0.96 to 1.15 IMN), the aircraft vibrated to such an extent as to preclude its use as a weapons delivery platform. The intensity of the vibration at a given Mach number increased with decreasing altitude; below 15,000 feet, the cockpit instruments vibrated to the point where they almost completely blurred.

The evaluation found the engine response poor. The engine accelerated slowly even at high power settings enough that its poor engine response precluded precise formation flying.

It also found the cockpit design antiquated, impossible to enter with any degree of urgency because of the time-consuming tasks associated with donning the parachute harness and hooking up the necessary personnel leads. The pilots rated the forward visibility as poor and found inconsistent labeling of the switches in the cockpit; the labels on the right side above the switches and the labels on the left side below the switches.

The MiG-21 looked very like an A-4 aircraft in size and shape. All pilots commented on the difficulty in determining one from the other when viewing both aircraft in the same area.

The A-4F, A-6A, A-7A aircraft possessed sufficient maneuverability in an initial break turn to thwart a MiG-21 attack. During every engagement where the MiG-21 attacked, fixed-wing, heavier than air attack planes (VA) (attack) forced the MiG-21 into a high yo-yo maneuver or overshoot. VA aircraft executing the break turn lost considerable energy and g available, allowing the MiG-21 to reattack with comparative ease if the VA aircraft elected to remain in the fight at slower speeds than the MiG-21.

VA aircraft had no control over the MiG-21's ability to disengage at any time throughout an engagement. The MiG-21's higher thrust to weight ratio allowed the MiG-21 pilot the option of continuing the engagement or disengaging at his discretion. The MiG pilots forced the VA aircraft to counter his offensive maneuvers if the planes remained engaged.

A-4F and A-6A aircraft demonstrated an ability to reverse and obtain a quick snapshot if the MiG-21 overshot close in at a high TCA (50° or more). When the MiG-21 overshot close in, the A-4 and A-6 used a maximum performance rudder reversal to attain a position behind the MiG-21. This placed them within the AIM-9D launch envelope with speeds for the A-4 and A-6 very low after completing this maneuver (about 130 KIAS). The A-7A lacked having available thrust to perform this high nose maneuver.

Comparisons between the F-4 and the MiG-21 indicated them evenly matched on the surface. However, air combat was not just about technology. In the final analysis, it was the skill of the man in the cockpit. The HAVE DOUGHNUT tests showed this most strongly. When the Navy or Air Force pilots flew the MiG-21, the results were a draw; the F-4 would win some fights, the MiG-21 would win others. There were no clear advantages.

The problem was not with the planes, but with the pilots flying them. The pilots would not fly either plane to its limits.

One of the Navy pilots was Marland W. "Doc" Townsend, then commander of VF-121, the F-4 training squadron at NAS Miramar. He was an engineer and a Korean War veteran and had flown almost every Navy aircraft. When he flew against the MiG-21, he would outmaneuver it every time. The Air Force pilots would not go vertical in the MiG-21.

The HAVE DOUGHNUT project officer was Tom Cassidy, a pilot with VX-4, the Navy's Air Development Squadron at Point Mugu. He had been watching as Townsend "waxed" the Air Force MiG-21 pilots. Cassidy climbed into the MiG-21 and went up against Townsend's F-4. This time the result was far different.

Cassidy was willing to fight in the vertical, flying the plane to the point where it was buffeting, just above the stall. Cassidy was able to get on the F-4's tail. After the flight, they realized the MiG-21 turned better than the F-4 at lower speeds. The key was for the F-4 to keep its speed up.

What had happened in the sky above Groom Lake was remarkable. An F-4 had defeated the MiG-21; found the weakness of the Soviet plane. Further test flights confirmed what was learned. It was also clear that the MiG-21 was a formidable enemy. United States pilots would have to fly much better than they had been to beat it. This would require a special school to teach advanced air combat techniques

Chapter 6 - MiG-21 Exploitation Results

The Air Force Flight Test Center completed a ground test, performance, and stability evaluation of the MiG-21 F-13 Fishbed-E, finding it a simple and highly reliable Mach 2 aircraft

During the technical exploitation, the exploitation team members found the MiG-21F-13 cockpit reflected the Soviet philosophy of engineering simplicity. Poor functional grouping of switches, controls, instruments, and warning lights gave the cockpit a cluttered appearance. The poor grouping of switches and controls required close pilot attention when requiring some cockpit action. However, overall design simplicity of aircraft systems required little pilot monitoring or control.

No major maintenance malfunctions occurred during the 102 MiG-21 flights. The team changed the tires, wheel brakes, and engine oil filter after approximately 50 flights. Three minor engine EGT (Exhaust Gas Temperature) system problems occurred. The team noted one small hydraulic leak. They experienced some difficulty during the initial flights with landing gear retraction; however, this did not affect aircraft availability. The canopy operating system was potentially weak. However, careful actuation of the canopy control minimized failure.

The MiG-21 required minimal ground-handling equipment for over-the-wing refueling and re-servicing of gasoline, lubricants, gaseous oxygen, and high-pressure air. They found the filler ports and access panels readily accessible and a 30-minute aircraft turn-around time frequently. Instead of using battery starts, the exploitation team normally used an external power source during this evaluation.

The team found the MiG-21 servicing requirements minimal. A crew of six men servicing and maintaining the MiG-21 often completed servicing between flights in 30 minutes without difficulty. They found all servicing receptacles readily accessible through individual access panels.

The exploitation team found the MiG-21 systems unsophisticated and designed for high reliability with no complicated servicing equipment required.

The MiG-21 utilized a self-contained electrical starting unit.

The main and booster hydraulic systems pressurized to a maximum working pressure of 3100 psi. It normally operated at 1200 to 1400 psi as compared to constant high pressures in US aircraft. The pneumatic system was a ground charged, a highly reliable system rated at 1800 psi. The fuel system was gravity filled and pressurized by sixth stage engine compressor air.

One boost pump supplied fuel directly to the fuel control. The 28 V electrical system contained a battery, starter-generator, and an inverter. Emergency electrical power was available from two batteries.

The MiG-21 was corrosion free. The skin's clear lacquer coating of the aircraft did not crack, peel, or deteriorate. The FTD, US AFSC, placed this substance under analysis.

The MiG-21 cockpit noise level was low. Compressor air and heat exchanging provided cockpit pressurization and air conditioning. The automatic cockpit temperature control preset before takeoff could not reset in flight. However, manual air temperature control was available.

No fog or snow blew into the cockpit through the system on any flight. The cockpit noise level was much lower than the F-4 or F-8 aircraft

The MiG-21 pilot seat positioning appeared to enhance pilot g tolerance. In the cockpit, the pilot sat with knees raised and his legs pointed more forward than down. Thus, the pilot's g tolerance appeared raised approximately one g.

The MiG-21 had poor cockpit visibility. The combination of a bulletproof glass plate, the gunsight combining glass, and the canopy restricted visibility through the forward windscreen to 3 to 5 nautical miles against F-4/F-8 sized targets. Visibility through the forward side panels and the remainder of the forward-hinged, clamshell canopy restricted the pilot's head movements, resulting in a 50-degree blind cone to the rear. The canopy rails stood much higher than in US aircraft and limited look down at the three and 9 o'clock positions to 20 degrees. "S" turning and looking through the forward side panels obtained excellent forward visibility in the MiG-21.

The project participants classed the MiG-21 cockpit layout and seat mechanization as generally poor, requiring a ladder to gain access to the cockpit.

Before pilot entry, maintenance personnel manually adjusted rudder bars. The pilot stepped on the seat, which contained the parachute; supported himself on the canopy rails; and carefully positioned his feet on the rudder bars. He then lowered himself into the seat. The pilot took great care to position his feet on the rudder bars because of the limited space between the leg restraint mechanism, center pedestal, and the lower instrument panel.

Seat comfort was marginal because of the parachute harness back strap arrangement, which the exploitation team alleviated somewhat on some flights by putting a foam rubber cushion between the harness and the pilot's back. The legs and buttocks positioned on the same level, which reduced the tendency for blood to pool in the lower body areas during g-forces application. The pilot accomplished up and down seat adjustment by an electric actuator. Canopy/head clearance restricted head movement. Seat adjustment accomplished by an electrical actuator, which moved the seat up and down. The seat positioning optimized pilot body posture, so the pilot is more easily tolerated high g loads.

The canopy was pneumatically operated by controls within the cockpit and externally accessible from the left forward nose section. The pilot positioned two levers in the cockpit to close and lock the canopy. There was no warning light to indicate a canopy-unlocked condition, requiring care when opening the canopy so as not to apply pneumatic pressure to the actuator before the locking mechanism had fully released. On one occasion, improper opening technique caused the canopy to snap open forcefully and disengage at the forward hinge point.

The MiG-21 required closing the canopy when taxiing. Limited over-the-nose vision and reduced acuity through the forward windscreen resulted in poor visibility when tracking. The narrow canopy restricted head movement. Ejection triggered on each armrest appeared easy to operate and readily accessible.

Donning the parachute and integral seat restraint harness required one to two minutes. Each leg strap on the seat-type parachute positioned over the leg and threaded through a harness loop and seat pan slot at the rear of the seat, then into a central harness connector. Finally, the pilot snapped the right shoulder strap into the connector and attached the oxygen, anti-g suit, and communication leads. The personnel lead group, although bulky, did not restrict pilot movement or cause discomfort. A ratchet handle located on the right side of the seat allowed the pilot to tighten the harness and restraint mechanism to a high tension. A release/locking lever located on the left side of the pilot seat adjusted the shoulder harness slack.

To the American pilots, the MiG-21 cockpit switches seemed considered poorly located. With slight slack in the shoulder harness, the pilot could actuate all switches and controls. With the shoulder harness locked in the fully retracted position, the pilot had some difficulty reaching the forward left and right extremities; i.e., landing gear panel and indicator light dimmer control.

Identifying placards positioned above each switch located the switches on the right vertical console inconsistency confused inexperienced MiG-21 pilots and caused identification difficulty.

They found good guards and covers for switches and buttons. However, the armament switches, controls, and monitoring lights for bombs, rockets, cannon, and missiles appeared located at random throughout the cockpit. Despite this scattered switch location, it required very little pilot action to set up the desired armament. When converting from a missile to cannon attack, the pilot repositioned the Missile—Cannon switch to "cannon," and uncage the sight cage lever, the latter accomplished by alternate use of the electrical cage function.

The pilots found the MiG-21's instruments poorly grouped and located. Pilot crosscheck required total panel scan instead of localized scanning. The instrument panel contained Mach meter, vertical speed, and turn indicators on the right half of the instrument panel. The left panel contained the attitude indicator, airspeed, altimeter, and compass. Engine instrument grouping was good. The engine monitoring gauges (tachometer, EGT, oil pressure, and fuel totalizer) located on the right lower half of the instrument panel. Overall, the pilots found good readability and interpretation of these instruments.

The exploitation team found the warning lights poorly located and difficult to interpret. They found the landing gear warning lights on the lower left sub-panel, and the marker beacon, nose cone position indicator, "Stabilizer ratio set for land" light, and trim warning placards in the center warning panel. The upper right portion of the instrument panel contained the fire warning and other lights. Dimness of the warning lights, even at full intensity, caused interpretation difficulty. They found inconsistent color-coding throughout the warning/monitor indicators, and red-colored warning light may or may not have been a normal condition. The monitoring and warning light system proved adequate for providing vital information to the pilot.

Manual control of the nose cone, stabilizer ratio, and intake shutter doors provided pilot override capability for these normally automatic systems. Emergency air start and landing gear controls proved adequate, however, required concentrated effort to actuate. The exploitation team incorporated an emergency hydraulic pumping unit for limited stabilizer control during primary and boost pump failure. This system automatically actuated or manually by the pilot. Aileron control effected manually if the booster system failed.

The control stick grip contained the speed brake, gunsight electrical cage, and trim armament fire buttons. Actuation of electrical cage felt somewhat awkward when pressing the trigger, however, did not necessarily limit the pilot's ability to operate the systems. The trigger normally remained stowed in an upright position and unfolded for operation. They found the brake handle arrangement poor and of antiquated design,

Throttle controls rated good to fair. The positive lock lever for idle was good since inadvertent stop cooking of the engine was nearly impossible. The afterburner engaging locking levers initially caused difficulty for the pilot because of the determined effort required to engage and disengage afterburner.

The MiG-21 appeared to have a high-speed ejection capability. The canopy designed semi-encapsulated the pilot during a normal ejection sequence. Alternate controls allowed for the separate jettison of the canopy. The ejection system was of unique design, partially encapsulating the pilot during the ejection. The seat formed an air blast shield to retain the front hinged canopy later jettisoned in the ejection sequence.

The MiG-21 ejection seats required several inputs such as pilot weight, height, etc., to dial in the proper barometric setting. By semi-encapsulating the pilot with the canopy during ejection, high-speed bailouts were possible without serious pilot injury. The system design operated at speeds up to 595 KIAS at sea level and up to 2.05 IMN at altitude. From all indications, this ejection system was extremely effective and reliable.

Armor plating protected the MiG-21 pilot as indicated below:
- Headrest .68 inches thick
- Rear Plate .63 inches thick
- Front Plate .4 inches thick
- Glass Shield 2.5 inches thick

Review of all available combat gun camera film indicated the MiG-21 tending to explode when hit by cannon/missile fire (probably due to wet wing design). However, in most cases, the pilot ejected successfully. The effectiveness of this armor plating apparently contributes to the high pilot survivability rate.

The MiG-21 design incorporated a three-wheel braking system. The nose wheel brake selected at the pilot's option, increasing the total system braking energy by 20 percent. After landing gear retraction, an automatic feature applied the wheel brakes to prevent tire rotation in the wheel wells.

During the exploitation team's verification of the MiG-21 gunsight radar capabilities, the gun mode obtained a 3–7 nautical mile maximum detection range in the missile mode and a 1.6 nautical mile maximum detection range.

The MiG-21 gunsight was ineffective during maneuvering flight. Manual ranging of the gunsight was not smooth or precise. System hysteresis and friction made it virtually impossible to prevent over control of the sight reticle diameter size with the throttle twist grip. Pipper jitter during cannon firing was more than 20 mils.

Gyro drift when tracking air targets was excessive. At g loads greater than +2.5, the sight reticle drifted to a point near the bottom of the sight combining glass. At very high g loads, the sight reticle disappeared entirely. The sight electrical cage functional was sluggish and slow to respond.

During air-to-air tracking, the pilot found it necessary to hold the electrical cage button (on the stick grip) until radar lock on occurred. He found the electrical cage button poorly positioned and difficult to actuate when preparing to fire the cannon. Over-the-nose visibility restrictions limited the useful mil depression to 95 mils. Large lead angles during air-to-ground attacks with bombs, cannon, or rockets remained unavailable. The pilot found it impossible to depress the gunsight in the cannon mode of operation as required for the ground attack at long slant ranges.

Maintenance Discrepancies
- 12 February 1968   #1 Boost pump inoperative
- 24 February 1968   EGT Malfunction (harness frayed)
- 28 February 1968   Frayed brake cable
- 5 March 1968   Oil System (6 sorties lost)
- 11 March 1968   EGT Malfunction
- 27 March   EGT malfunction

The oil system did not malfunction. (Unfamiliarity with the aircraft made a clogged oil filter seem like a major problem)

Still, the exploitation effort lost only 11 sorties. The US jets did not come close to that.

Summary of technical Evaluation of MiG-21 Unique Design Features
- Very Low Wing Loading (50–55 psf)
- Lacquer Coating for Corrosion Prevention
- Ejection System (SK-1 seat and canopy)
- Air Intake (3-position, normal, Mach 1.5, Mach 1.9)
- Seat Position adjustable, however, cramped
- Low Maintenance Requirements
- No roll, pitch, yaw stability augmentation
- Cooled Navigation Lights
- Optimized Cross-section

Smooth surface finishes only where it needed to be. Rivets protruded in out of sight locations.

Summary of Tactical, Operational Applications

Project flights conducted to duplicate the Air Combat Maneuvering environment encountered in SE Factors contributing to test results that did not simulate this environment include:
- Participating pilots briefed on maneuver performance before each flight.
- Two-way UHF radio communication maintained between participating pilots.
- Engagements terminated when encountering unusual flight characteristics.
- Bingo fuel weights adhered to.

Life or death situation was not present; however, participating aircraft constantly flew to maximum performance. Only aircraft performance was evaluated since missiles and guns not fired during engagements.

The exploitation results recommended:
- The following general fighter tactics applied when engaging a MiG-21:
- Act aggressively. Using sound tactics while maneuvering for the advantage.
- Determining your opponent's ability—assuming the "Red Baron" engaged until proven otherwise.
- Using the Combat Spread Formation when flying a hostile area to provide visual coverage of each aircraft's stern are engaged as a section to provide mutual support. The MiG-21

was extremely difficult to see due to its small size. The MiG-21 was normally under GCI control and positioned for a stern area attack.

- Maintaining a minimum of 450 KIAS while patrolling allowed the instantaneous application of maximum g.
- Maintaining a high energy level while engaging. Trade airspeed for altitude only and never attempts slow speed scissors. If necessary, dive away to regain airspeed for a reattack or to execute an escape maneuver.
- Forcing the fight to low altitudes to take advantage of the MiG-21's airspeed limitations and high stick forces below 16,000 feet.
- Using lag pursuit maneuvering close in. Because of the MiG-21's superior turning performance, a close-in overshoot was highly probable if using lead pursuit to close for a minimum range missile or gunshot. As the MiG-21 initiates a defensive hard or break turn, maneuver to 3,000 to 5,000 feet astern and outside the MiG-21's radius of the turn. This prevents a close-in overshoot, reduces energy bleed-off, and places you in his blind cone. Continue maneuvering the stern area until it appears the MiG-21 losing visual contact, then close for the kill.
- Maneuvering into the MiG-21's blind cone during all offensive maneuvering to capitalize on the MiG-21's visibility restrictions and to arrive in the aft hemisphere missile envelope.
- Avoiding dissipating energy by using hit and run attacks and yo-yo high. Do not strive for a rapid close-in shot.
- Executing a high g, roll away to position for a lag pursuit attack instead of performing a high yo-yo to counter the overshoot if a close-in overshoot is imminent during offensive maneuvering. This will eliminate the possibility of a slow speed scissoring situation catching them.
- Using an oblique loop maneuver for reciprocal course changes below 16,000 feet once engaged; vice horizontal, high-yo-yo, or low yo-yo type turns. The oblique loop allowed the attacker to keep sight of the MiG-21 during the maneuver while capitalizing on the F-4/F-8's superior performance in the vertical plane, a maneuver repeatedly used as an effective position maneuver.
- When sighting a MiG-21, turn to engage head-on. Reduce lateral separation and jink to avoid cannon fire while closing. Turning to position the MiG-21 at a high TC if unable to engage head-on at that range, maintain this high TCA while accelerating for separation. Be prepared to break into an ATOLL missile or to negate a gun-firing pass and force an overshoot. If a wingman is present, he separates in the vertical to present two targets and employ loose deuce maneuvering to sandwich the attacker.
- Reducing lateral separation between aircraft to a minimum if committed to a head-on attack. A MiG-21 can convert any lateral separation into a decreasing TC
- When passing the MiG-21 head-on, the delay in the turn back into him took up to 5 sec or 90° of bogey turn. Allow the F-4/F-8 to accelerate and provide sufficient lateral separation again to meet the MiG-21 head-on after the reversal. A turn initiated immediately after passing losses energy and the MiG-21 gains TCA in turn. As subsequent head-on high-energy passes, continue the turning to dissipate the MiG-21's energy and make it vulnerable.
- When section maneuvering offensively to engage a MiG-21, close until the MiG-21 initiates a defensive maneuver. When the MiG-21 maneuvers, the wingman separates vertically. One aircraft keeps the MiG-21 engaged, while the other employs loose deuce maneuvers to close for the kill.
- If the MiG-21 closed to gun-tracking range (within 3,000 feet/30 degrees angle off), escape becomes difficult. Execute a low nose break into the attacker, accelerating to above 595 knots at maximum g for that speed; keeping the attacker in sight to affect an escape maneuver. If fired upon, vary the g load and yaw on the aircraft to negate tracking

solutions. If the attacker follows you down into the high q low altitude region, consider an engagement. If the attacker rides high, do not pull back up to re-engage. Separate and evaluate the situation.

Along with the tactics recommended above, the team found the following tactics pertinent to the F-4 weapons system.

A-visual retention of the MiG-21 beyond 2 miles was very difficult. AIM-7E/E-2 trigger squeeze minimum range thumb rules in the forward quarter were 3 miles and 2 miles respectively. Before firing an AIM-7 missile, radar lockup required 4 seconds to allow radar settling and missile speed gate tuning. Resultant minimum ranges to allow for radar lockup in the forward quarter were 4/3 miles for the AIM-7E/E-2 respectively.

The probability of visual detection of the MiG-21 at 3–4 miles in the forward quarter aspect proved remote. Successful simulated AIM-7 forward quarter firings resulting from an initial visual detection of the target did not occur. This did not include VID (visual identification) formation forward quarter simulated firings or CIC authorized firings.

Attacking the MiG-21 with 3–4 miles lateral separation made a successful simulated AIM-7E-2 shot possible during the turn back if accomplishing the lock-on before 45° to go. The time delay from radar lock-on with 45 degrees to go before roll out at 2 miles, head-on, approximates 4 seconds. This normally provided sufficient time to launch an AIM-7E-2 missile head-on. The rapidly changing bogey azimuth and elevation in this aspect required a high level of crew proficiency.

The F-4 two-man crew had a significant advantage over single-seat fighters. Once engaged, the RIO was available to and concentrated on keeping sight of the attacker/attackers until radar acquisition was possible. The RIO repeatedly proved invaluable to the success of the engagements. Padlock lookout was mandatory against aircraft as small as a MiG-21.

The energy advantage of the F-4 below 16,000 feet and above 450 KCAS allowed the F-4 to gain and maintain the offensive with the techniques of lag pursuit, vertical reversal, and employment of slash attacks.

If sighting the MiG-21 in the forward area closing, turn to meet him head-on with minimum lateral separation. If the MiG-21 did not turn to meet you head-on, assume he did not see you. Turn for lateral offset to convert to an aft hemisphere attack maintaining rigid lookout doctrine during the turn to. Avoid turning in front of a trailing wingman.

If the MiG-21 turns toward you before passing abeam, turn into him to reduce lateral separation. Jink as necessary to negate a head-on gun attack. While closing, radar search the area behind the MiG-21. As the MiG-21 passes close abeam, drop a wing as necessary to keep the MiG-21 in sight. Maneuver in a maximum performance oblique loop to re-engage.

Analyze the MiG-21 pilot's ability in this first turn. If he was not maneuvering aggressively, position yourself for a kill. If he maneuvered aggressively and had gained any advantage in the reversal, attempt again to meet him head-on. On passing, delay the turn back slightly, approximately 5 seconds (keeping the bogey in sight). This ensures sufficient lateral separation to compensate for turn radius and to ensure a subsequent head-on pass. If the MiG-21 maneuvered and aggressively reversed on each pass, his energy level dissipated.

If engaged above 16,000 feet force the fight to low altitude. Below this altitude, use the oblique loop to affect a turn reversal. The vertical reversal capitalizes on the F-4 energy advantage in the zoom and forces the MiG-21 to work the vertical and dissipate his energy.

The F-4 speed on top was normally 250–300 KCAS. The MiG-21 does not regain his energy in the dive as rapidly as the F-4. When the MiG-21's energy level dissipates, and you gain an advantage, continue pressing the attack, however, do not rush it. Capitalize on your performance and exploit the MiG-21's limitations. Continue to perform slashing attacks; employing yo-yos until achieving a missile launch position.

In the event, a close-in overshoot appeared imminent, execute a barrel roll, or roll off a maneuver to the MiG-21's blind area, approximately 1 mile aft in a lag pursuit attack. Retain your energy level. If a close-in overshoot develops, rapidly roll to effect separation, unload, keep the bogey in sight, and reengage on your terms.

When sighting a MiG-21 with separation available, meet him head-on, and continue as above. If a head-on meeting is impossible, turn to keep the MiG-21 in sight, place him at a high TCA and accelerate. As the bogey closed, be prepared to break into a missile. Continue to keep the MiG-21 in sight, place him at a high TCA until a break turn was necessary to negate a gun attack and force a high angle overshoot. The MiG-21 will most probably yo-yo high to conserve energy and to maintain the offensive.

If the MiG-21 overshot flat and slid out ahead, a reversal or roll over was possible to arrive at his 6 o'clock position. Using extreme caution if not achieving immediate success. Be prepared to unload into him and accelerate for separation immediately.

The MiG-21 was far superior at close in maneuvering than the F-4. Never attempt close in maneuvering unless it becomes apparent the MiG-21 pilot is incompetent or holding a definite position advantage.

The MiG-21 repeatedly demonstrates the ability to counter a high angle overshoot and rapidly regain a gun-tracking position, if the US fighter reverses into him as he overshot. Maneuver to his blind area achieved a missile launch position, and immediately regain your energy level. Be sure the MiG-21 was not a decoy feinting poor performance. A competent MiG-21 pilot turned an apparent defensive situation for him into a very sudden offensive position. If a scissor situation was imminent, dive into his blind area and accelerate for separation.

(KIAS) Keep the MiG-21 in sight. Obtain sufficient lateral separation to reverse back into the MiG-21 to affect a head-on engagement. Attempt to acquire the MiG-21 on the radar in the turn for a possible Sparrow shot. The RIO remained padlocked on the MiG-21 until it was well within radar gimbal limits (45 degrees and the pilot had confirmed visual contact). When the RIO goes to the radarscope, the pilot padlocked on the MiG-21 and coached the RIO on his position. As the MiG-21 approaches the nose, PLM may employ if the RIO was unsuccessful in acquiring a radar lock-on. Feel out the MiG-21 during subsequent maneuvering and effect a kill when and if presented the opportunity.

The exploitation established three rules as essential to successful F-4 section tactics against the MiG-21 or any small; low wing loaded aircraft They are:

All crew maintained visual contact with the bogey. The RIO could not return to the scope until the bogey was within approximately + 45 degrees of the nose and the pilot had confirmed visual contact (2) Engage only in section. It was very easy to split the F-4 section, lose mutual support, and fight two separate one-on-one engagements. In combat, when the possibility of multiple bogies existed, splitting the section could be disastrous.

A steady flow of information took place within the F-4 section, requiring the relaying relative position to each other, relative position of the bogey, intentions, and tactical orders.

Mutual support between F-4s engaging a MiG-21 dictated that each aircraft in the section protect and support the other during an attack. Each member of the section had sufficient, however, not excessive separation to launch missiles at any threat posed. Also, the F-4 section required positioning to prevent the bogey from working both F-4s as a unit, while maintaining contact with each other.

Apply maximum separation between F-4s-3 miles on VID formation, maintain maximum separation between F-4s abeam, co-heading one mile, and offset heading on the offensive about 90 degrees. Set up a two-on-one offensive attack that forced the bogey to meet threats from divergent angles.

After a head-on pass, the F-4 section maneuvered as necessary to maintain mutual support and visual contact.

AFFTC Lessons Learned

- Power checked at Mil power before brake release. Brakes failed to hold in afterburner
- Rudder effectiveness occurred at 45 knots
- Nose wheel liftoff at 114 KIAS (with full aft stick)

Exhaust smoke comparison

- At 15,400 pounds, with 30 degrees (full) flaps, takeoff speed was 165 KIAS
- Afterburner failed to light when selected until after military thrust achieved
- Stabilator was the only trimmable control surface
- The engine never stalled

ADC Comments on Project HAVE DOUGHNUT

ADC found visual contact difficult because of small size, except the silver color made the planform view relatively easy to see.

The radar signature to the MA-1 fire control system indicated contact and tracking adequate for Intercept completion with contact 20–25 miles in all aspects. ADC found stern contacts the best and head-on co-altitude the worst.

Regarding an F-4, it provided a return 1/2 as large in front, 3/4 in beam and almost identical in stern because of engine modulation. The IR signature to the MA-1 fire control system proved adequate for acquisition and tracking. ADC found it about 3/4 as strong a return as the F-4 and like the F-106 in military, however, not as strong as F-106 in A/B afterburner.

ADC found the performance of the test aircraft not as good as expected. Limiting factors making it capable of exploitation included slow engine response in military required for A/B initiation, q limit below 15,000-feet, visibility to the rear, and visibility over the nose.

Another area needing exploiting included energy and time required to transit the transonic zone with pylons and missiles aboard. Its controls stiffened at low speed and low altitude. It lacked endurance at maximum power and had only one radar fire control system.

ADC found the F-106 aircraft's radar capable of acquisition and usable to put the F-106 in position for armament launch. The F-106 used radar snap-up attacks with all-aspect armament load to exploit test the MiG-21's lack of adequate fire control system. The F-106 took advantage of the inability of the MiG pilot to see over the nose and through the windscreen.

The ADC recommended using the F-106's ability to accelerate faster than the MiG-21, enabling it to achieve a higher speed (beyond q limit) to separate anytime it found the F-106 not in an advantageous position during engagements. The F-106 could use missile launch and then use lag pursuit while closing to gun kill position, depending upon its superior turn capability to pull necessary lead for gun firing.

ADC recommended expediting procurement of cannon for F-106 for near-term close-in-kill armament and use the superior zoom capability of the F-106 to advantage for repositioning after separation during engagements.

The ADC cautioned the bar overhead in the F-106 canopy causing F-106 aircrews to lose sight of the MiG-21 during close-in engagements. With this defect, rolling maneuvers could get the F-106 into trouble and every effort made to expedite replacement of the F-106 canopy bar with a clear pane.

Following the exploitation, ADC cautioned F-106 aircrews to take care to preclude unnecessary expenditure of energy when observing the MiG-21 initiating a turn. The appearance of a generation of a great amount of turn was deceiving.

The misleading size of the MiG-21 caused an error in estimation of range and rate of closure by F-106 aircrews. Thus, the F-106 should not attempt slow speed turning contests with the MiG-21 as the performance was close to equal, and a slight miscalculation could be fatal. The F-106 should keep its speed at 400 to 450 KCAS during patrol and engagement.

The exploitation determined the F-106 could use its missile armament during an engagement with efforts to modify the fire control system with IE "boresight expedited. The IR "boresight modification should include an automatic radar look-on by caging the radar antenna to the IR head. It should sweep the range gate out to affect radar look-on with the option of caging the radar antenna dead ahead.

The ADC exploitation of the MiG-21 found the F-106 patrol formation, as taught at the Interceptor Weapons School, provided adequate protection against surprise attack by the test aircraft. It recommended improving F-106 radar, and all flight members know visual search patterns to insure responsibilities for sector search.

The evaluation determined radar tracking procedures are lacking and the need for procedures for assigning responsibility within and between elements for armament launch when some or all members of the flight acquire a MiG-21. The F-106 aircrews needed training that when pressing an attack after acquiring the test aircraft, they expended armament in order of priority, i.e., missiles, then press to gun position or separate if no gun is aboard or all missiles expended.

The ADC found an all-aspect armament capability with a successful qualification of 5 out of 5 missile simulator evaluators (WSEMS) during the test. This indicated that properly prepared AIM-4F missiles would see the test aircraft as far out as 3.5 miles on front aspects with a high probability of successful guidance. The F-106, with its present configuration and with developing tactics, proved capable as an effective counter to the MiG-21.

Findings: The Bottom Line
Simplicity; Ease of Flying–It is a good, honest aircraft
Reliability and Maintainability (20-minute turn around)

Cross-Sectional Area

Engine exhaust smoke was another exploitable discovery. The Fishbed's afterburner marked the aircraft's location by producing white puffs of unburned fuel when it engaged or disengaged, a small consultation because the American jet fighters similarly aided the enemy by leaving a smoke trail that exposed their presence.

3-wheel Brake Concept.

Armament was adequate for an interceptor. For the point interceptor role, the MiG-21's basic weapons included a 30-mm cannon loaded with 60 rounds of ammunition and 2 AA-2 ATOLL heat-seeking missiles. The Soviet-built ATOLL missiles copied the US-made AIM-9 Sidewinder obtained and reverse-engineered by the USSR when one lodged in a MiG-17F and failed to explode.

In an air-to-ground role, the MiG-21 carried the 30-mm cannon and two pods containing 32 57 mm folding-fin aerial rockets. The cannon proved potentially lethal against tanks, however, encountered considerable piper (gunsight) jitter while strafing. Another drawback for the MiG-21 was its high-speed-low altitude stability in rough air.

Excessive bleed-off during high-g turns turned into an advantage for the MiG by decreasing its turn radius while sustaining its g-force at slower speeds than comparable US fighters. Obviously, in a turning fight, this gave the Fishbed a tactical advantage.

HAVE DOUGHNUT confirmed American fighters flew faster than the MiG in actual combat in the skies over North Vietnam. The MiG pilots overcame this US advantage by tactfully drawing the US fighters into turning engagements where the superior speed did not matter. The MiG knew the appropriate angle to cut their circle and make the MiG guns effective. Maintaining high-speed at low altitude was the key to MiG survival.

Under the HAVE DOUGHNUT and HAVE DRILL programs, the exploitation team used the first

MiG flown in the United States to evaluate the aircraft in performance and technical capabilities. The team also evaluated the MiG-17 in operational capability, pitting the types against US fighters.

Meanwhile, US Air Force and US Navy airmen flying contemporary advanced aircraft, combined with a legacy of successes from World War II and the Korean War, revamped aerial combat because of Project HAVE DOUGHNUT at Area 51. With what they learned at Groom Lake, the Navy put designs on the drawing board for an entire generation of aircraft. The engineering for this aircraft optimized daylight air-to-air combat (dog-fighting) against both older, as well as for emerging MiG fighters.

The Navy created the TOP GUN Weapons School in 1969 and experienced strong results against the MiG-21 when they encountered it in 1972.

Putting Guns Back on the US Fighter Planes

In 1968, Chief of Naval Operations (CNO) Admiral Thomas Hinman Moorer ordered Captain Frank Ault to research the failings of the US air-to-air missiles used in combat in the skies over North Vietnam, this bringing Operation Rolling Thunder.

Operation Rolling Thunder lasted from 2 March 1965 to 1 November 1968 and ultimately saw almost 1,000 US aircraft losses in about one million sorties. The operation became the Rorschach test for the Navy and Air Force, which drew nearly opposite conclusions.

The USAF concluded its air losses primarily due to unobserved MiG attacks from the rear. The Air Force treated it as a technology problem and responded by upgrading its F-4 Phantom II fleet by installing an internal 20 mm Vulcan cannon. It replaced the gun pods carried under the aircraft's belly. It developed improved airborne radar systems and worked to solve the targeting problems of the AIM-9 and AIM-7 air-to-air missiles.

Thus, the Air Force did not create a dissimilar air combat program until 1972/73 when the Air Force created RED FLAG to give its pilots a better edge in the fight. The USAF finished the war in 1973 with a two-to-one kill-loss ratio, downing 137 MiG while losing 65 aircraft, including bombers, to MiGs.

- A g-load factor was 8 g without stores, 6 with stores.
- Max indicated airspeed-595 knots below 15,000 feet 640 knots above 15,000 feet;
- Maximum indicated Mach _ 2.05 without stores, 1.6 with stores.
- Strike radius-370 nautical miles with external fuel.
- Poor forward and rearward visibility F-4 acquired in 3–5-mile range.
- Low Q Limit-Below 15,000 limited to .98 Mach or 595 knots–severe buffet.
- Afterburner puff-Above 15,000 Fishbed-E produced a puff in/out of afterburner
- Engine response-extremely slow.
- Cockpit noise-extremely low.
- Gunsight capabilities-3.7 Nautical Miles, missile mode: 1.6 Nautical Miles, gun. Gun/missile target is tracking impossible over 3 g.
- Slow speed The MiG-21 could maneuver at 115 KIAS.
- The MiG-21 engine was technologically behind its US counterparts, so spool-up from idle to the full military power required 14 seconds, with a tendency to hang up in the process. This often led to hot compressor stall or engine over-temperature conditions.
- Low Altitude Transonic Vibration. Typical of delta wing aircraft, excessive high g turns caused airspeed bleed-off.
- Formation flying proved difficult because of the mix slow engine spool-up handicap, its taking 14-seconds to accelerate from idle to full power and formation maneuvers required constant use of speed brakes and rapid throttle movement.
- Flying in Turbulence caused poor directional stability.
- Night Flying.
- Gunsight was deficient. The tracking index drifted off the bottom of the windscreen when tracking targets exceeding three G.

• Exceptionally heavy pitch force required above 685 mph or a .98 indicated Mach Number.

• Problematic recovery during dive-bombing, strafing, or air-to-ground rocket is firing.

• Exceptionally slow engine acceleration from idle to full military power.

Design to prevent over stress problems during the pull up from a target made it difficult or impossible to achieve high pitch rates expected of a fighter-bomber. Heavy pitch forces at high-speed limited the pilot's ability to recover from a diving attack or maneuver while approaching and departing the target.

Most significantly was the aircraft's inability to go supersonic below 15,000 feet. Severe buffeting at low-level prevented it from exceeding 685 mph or .98 Mach airspeeds. This airspeed limitation exposed a major exploitable design flaw that F-4s and F-105s exploited during the Vietnam War by typically approaching the MiG-21s at 633 mph then departing supersonic.

Easy to kill non-sealing tanks, unprotected engine, light metal structure, high-pressure O2 bottles, 85% kill probability.

A special limitation for the day visual was the fighter-interceptors front and rear visibility. The combination of a bulletproof glass slab and the windshield restricted forward visibility through the gunsight. The protective seat flap over the pilot's head and the narrow design of the ship's canopy and fuselage structure handicapped visibility in the 50-degree tail cone.

What the Planes Should Do If Encountering a MiG-21

- If you are flying an F4C, D, or E and a MiG 21 finds you below an altitude of 15,000 feet you had better run away. Speed is life.

- If you are flying an F-105 and a MiG 21 finds you, avoid prolonged maneuvering engagements. Get into the MiG-21 blind cone, use mutual flight support, and use hit-and-run tactics. Maintain maximum airspeed below 15,000 feet and do not get in a turning fight. Speed is life, and you can only go in a straight line.

- If you are flying an F-111 and a MiG 21 finds you, remember the MiG has superior turn performance everywhere. Below 15,000 feet, the MiG-21 has superior level acceleration up to its speed limit. When attacked get below 15,000 feet and above the MiG-21 speed limit of .98 Mach. Again, speed is life, and you can only go in a straight line.

- If you are flying an F-100D and a MiG 21 finds you, remember the MiG-21 has significant turn performance and level acceleration advantages. Avoid maneuvering, use hit-and-run, mutual support, and again, Speed is life, and you can only go in a straight line.

- If you are flying an F-104 and a MiG 21 finds you, remember the MiG-21 has superior turn capability. Avoid maneuvering, use hit-and-run, mutual support, and defend yourself by accelerating above MiG-21 limit speed. Speed is life, and you can only go in a straight line.

- If you are flying an F-5A, you can closely simulate the MiG-21 up to Mach 1.2 for combat engagements.

- If you are flying an RF-101 and a MiG 21 finds you, remember the MiG-21 has a slight advantage in afterburner and superior turn capability everywhere. When caught by a MiG-21 dive and run, speed is life.

- If you are flying a B-66 and a MiG 21 finds you, you are a target.

- If you are flying an RF-4C and a MiG 21 finds you, dive, and run, speed was life

- If you are flying a Navy F-8E and a MiG 21 finds you, the MiG can out turn the F-8 in a close in fight. Speed if life.

- If you are flying at Navy A-4F, A-6A, or A-7A and a MiG 21 finds you; the MiG-21 pilot can choose to engage or disengage you at will. You are a target, however, with a chance to live.

Text HAVE FERRY MiG-17F FRESCO

Crashed at Tonopah, Nevada Project Constant Peg

HAVE DRILL/HAVE FERRY TACTICAL

TAC Evaluation
Performance Comparison

AREA OF COMPARISON	F-4	F-105	F-100	F-5
TURN				
0.9 Mach/450 KIAS, MAX	Comparable	Superior	Superior	Superior
0.9 Mach/450 KIAS, MIL	Comparable	Comparable	Comparable	Superior
350 KIAS, MAX	Inferior	Inferior	Inferior	Comparable
350 KIAS, MIL	Inferior	Inferior	Inferior	Comparable
ROLL				
450 KIAS	Superior	Superior	Superior	Far Superior
350 KIAS	Superior	Superior	Superior	Superior
SPEEDBRAKE DECELERATION				
Constant Power	Inferior	Inferior	Inferior	Inferior
Idle Power	Inferior	Inferior	Inferior	Inferior

AREA OF COMPARISON	F-4	F-105	F-100	F-5
ACCELERATION				
Level, MIL	Superior	Superior	Superior	Superior
Level, MAX	Superior	Superior	Superior	Superior
Level, MIL to MAX	Inferior	Inferior	N/A	N/A
Unloaded, MIL	Superior	Superior	Comparable	Slightly Superior
Unloaded, MAX	Superior	Superior	Superior	Superior
Unloaded, MIL to MAX	Inferior	Slightly Inferior	Slightly Inferior	N/A
ZOOM				
MAX to MAX	Superior	Superior	Comparable	Superior
MIL to MIL	Superior	Superior	Comparable	Superior
MIL to MAX	Comparable	Comparable	N/A	N/A

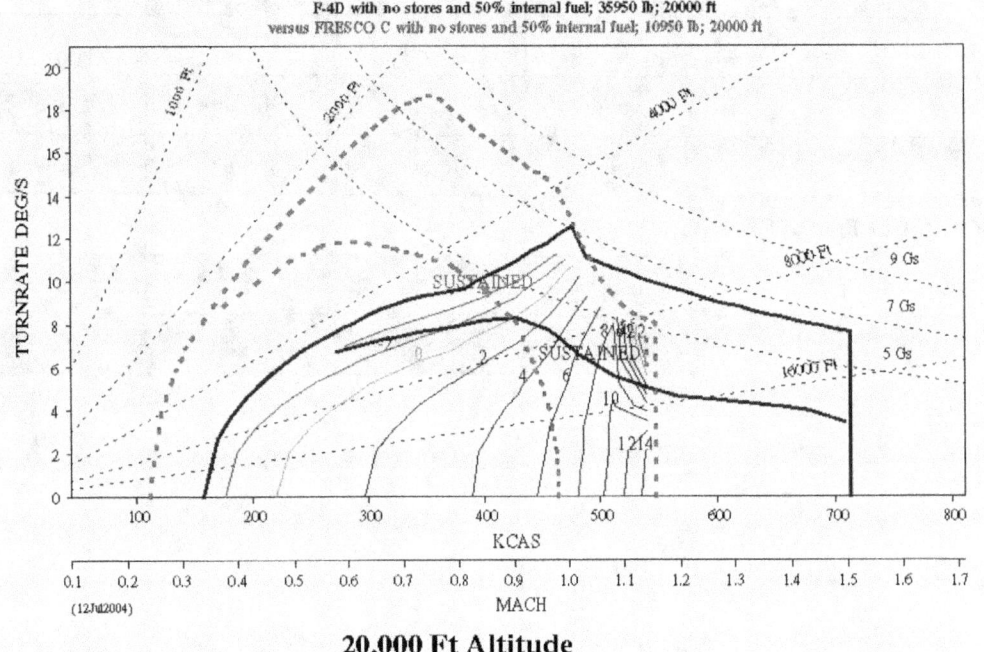

F-4D vs. FRESCO C
Delta Longitudinal Acceleration (KCAS/sec)
(Turnrate vs. Mach)
Maximum Afterburning Power

Contours represent F-4D performance minus FRESCO C performance.
Solid boundary is F-4D envelope.
Dashed boundary is FRESCO C envelope.

F-4D with no stores and 50% internal fuel; 35950 lb; 20000 ft
versus FRESCO C with no stores and 50% internal fuel; 10950 lb; 20000 ft

20,000 Ft Altitude

Background of the MiG-17F Technical Exploitation

Following the successful exploitation of the HAVE DOUGHNUT MiG-21 Fishbed a year earlier, considerable interest focused on an even more intense exploitation of the older MiG. The MiG-17F was widely exported and operationally deployed to far eastern nations within the Communist sphere.

During the Vietnam War, the cannons of the MiG-17F continued to dominate the air war against the superior aerial assets of the US. This limited war environment made the exploitation of Communist operational systems a vital necessity. The FTD at Wright-Patterson recognized its basic mission included the evaluation of foreign material of scientific and technical value to both the national intelligence community and Air Force research and development organizations.

US aerial combat losses over Southeast Asia further stressed the MiG-17F threat and demonstrated the need for a tactical as well as technical evaluation of this weapon system. Complete volumes of handbooks and specifications remained available in the intelligence community on the MiG-17. Nonetheless, the exploitation team needed to evaluate the actual aircraft to obtain vital tactical and operational information necessary for the effectiveness and survival of their air warfare exploitation teams. At the same time, exploiting an operational aircraft provided the opportunity to reevaluate current hand-book-data. The technical personnel gained an excellent insight to past levels of Soviet technology to allow their more clearly predicting future advances. Likewise, the exploitation of current threat aircraft such as the MiG-17F was significant. Its research and development application revealed the technological effort for future developing and testing counter-threat systems.

The HAVE DOUGHNUT special projects veterans based at Area 51 recognized the need for further exploiting the aerial assets of their enemy and honed their skills and equipment to provide this service to their customers. Customers for this project included the Navy, USAF, and Intelligence customers, i.e., the FTD of AFSC (a.k.a. National Air & Space Intelligence Center [NASIC], AFFTC, Air Defense Command [ADC], SAC, Aeronautical Systems Division [ASD], Tactical Air Command [TAC], and Navy Weapons Center [NWC]. The mutual need for acquiring a MiG-17F ended four months after the conclusion of HAVE DOUGHNUT. Two Syrian MiG-17Fs lost on a navigation exercise inadvertently landed at an Israeli air base. On August 12, 1968, while flying two MiG-17F Frescos, disoriented Syrian first lieutenants Walid Adham and Radfan landed in Israel by mistake. These were Polish-built Lims. (The United States would later acquire a number of Polish MiG-15 and 17 Lims directly out of the Polish air force's storage facility at Katoswice-Mierzecice.)

This acquisition of the slow and dated MiG-17s was of high importance to the United States as they represented the main adversary type encountered in the skies of Vietnam. Though limited to subsonic performance, VNAF MiG-17s were out flying American fighter pilots, this leading to dismal kill ratios. After testing, the Israelis turned the two MiG-17s over to the United States for analysis. In 1969, the exploitation team already assembled at Area 51 added the two ex-Iraqi MiG-17s transferred from Israeli stocks to the operation under the project names HAVE DRILL and HAVE FERRY. The Air Force assigned US designations and fake serial numbers to identify them in DOD standard flight logs.

The Project HAVE DRILL code name identified a joint Air Force/Navy program, a program to evaluate and exploit the capabilities of MiG-17F aircraft and its power plant, the VK-1F engine, both developed in the early 1950s. Initially produced by other Communist nations. MiG-17F as a "day fighter," the MiG-17F was still used extensively.

The FTD of the Air Force managed the overall project with the Naval Scientific Technical Intelligence Center coordinating the Navy's effort.

FTD maintenance performed on the MiG-17F HAVE DRILL and HAD FERRY MiG-17s began with the preflight and technical phase of Project HAVE DRILL.

The HAVE DRILL aircraft arrived at the Area 51 exploitation site on 27 January 1969. Uncrating and assembly began on 29 January 1969. During assembly, FTD examined the aircraft for discrepancies and after assembly performed a 50-hour inspection. FTD then repaired, adjusted, and operationally checked during the inspection the noted aircraft discrepancies to render the MiG-17F ready for flight in 19 days.

- Discrepancies Found During Technical Exploitation.
- Right-hand brake air line damaged beyond repair,
- Right-hand main gear door actuator hydraulic hoses deteriorated.
- Left-hand main gear door actuator hydraulic hoses deteriorated,

- Left horizontal elevator damaged at trailing edge,
- SRD-IM Radar inoperative.
- SIRENA coax cable to the antenna leads broken.
- SIRENA-2 unit #3 cannon plug loose and not safetied.
- APT fuel tank transfer fuel line is leaking aft of quick disconnect.
- APT fuel tank vent line is leaking APT of the quick disconnect.
- Oxygen line in nose bay leaking,
- HF inverter removed for troubleshooting and adjustment,
- Seat ejection cartridge outdated,
- Air line on gun package damaged.
- Speed brake selector valve leaking.
- Hydraulic leak in left wing at the fuselage.

During the assembly and 50-hour inspection, FTD personnel inspected, adjusted, repaired, and operationally checked all the systems while installing or altering the instrumentation necessary to exploit the technical aspects of the plane.

Description of the MiG-17 Tests Item

Of the four operational variants of the Fresco, the exploitation team confined this description to the Fresco C version used in the conduct of this test. The following physical dimensions apply to that aircraft

- Wingspan 31.5 feet.
- Length 38.32 feet.
- Height 12.47 feet. 23 mm
- Diameter of fuselage 4.76 feet maximum.
- Takeoff gross weight (clean) 11,803 pounds.
- Takeoff gross weight with external fuel: 13,414 pounds.

The sweep of the wings varied from 49 degrees inboard to 45 degrees outboard. The horizontal tail swept back 45 degrees, while the vertical portion swept 55 degrees, 41 minutes.

The VK-1F engine powered the Fresco C, an afterburning version of the earlier VK-1. The engine was a centrifugal flow compressor type, with a two-position nozzle positioned full-open for maximum power operation, and full closed for military power. The thrust rating varied from 5,720 pounds at military power to 7,440 pounds at maximum power.

One 37 mm internal cannon and two 23 mm internal cannons, all nose-mounted, comprised the primary armament of the Fresco. The 37-mm cannon had a capacity of 40 rounds, while the two 23 mm cannons hold 80 rounds each. Total continuous firing time available with a full load of ammunition approximated six-seconds. An optical gyroscopic lead computing sight, using a range only radar, provided the pilot with a solution to the firing problem. The exploitation team knew from observation that Fresco C carried two externally mounted general-purpose bombs instead of external fuel tanks when serving in a tactical role.

Simple. However, adequate, avionics characterized the Fresco C. The range only radar defined required range information for the pilot The Fresco had no radar on board. It had the common X-band passive tail warning radar receiver, designated Sirena with a nominal range of five nautical miles when illuminated by a peak power source. The inclusion of a SHO-2 L-band transponder permitted an Identification Friend or Foe (IFF) capability.

The purpose of this test determined the best USAF tactical aircraft offensive and defensive maneuvers against the Fresco C (MiG-17F) through flight test to verify/modify data and conclusions contained in tactical doctrine.

The MiG-17F Pilots

- USAF Col Robert `Bobby' Bond (who later flew A-7Es in Southeast Asia),
- Navy Commander Marland W 'Doc' Townsend (an F-4 Phantom II pilot),
- Navy Commander Tom Cassidy (a future admiral), and
- Navy Commander Foster S 'Tooter' Teague (who later commanded a carrier air wing in Vietnam).

The two MiG-17s flew 198 sorties (usually together) against a variety of US Navy warplanes, ranging from the F-8J Crusader to the A-6A Intruder. Later, separately, they flew against the USAF F-102A Delta Dagger, F-104A Starfighter, and F-106A Delta Dart.

The Shock from Exploring the MiG-17

The HAVE DRILL and HAVE FERRY evaluation details of testing two MiG-17s were minuscule. This caused some to question the reason for testing this older vintage aircraft from the Korean War era. This was especially so since volumes of handbooks and specifications already existed on the MiG-17F in the intelligence community.

This line of thought changed somewhat when pilots found how the canopy-mounted periscope in the MiG-17F improved visibility to the rear hemisphere.

This and other hands-on observations convinced the Air Force of evaluation of the actual aircraft furnishing vital tactical and operational information necessary for the effectiveness and survival of their air warfare exploitation teams.

The shock and wisdom of this decision to go with the program came to fruit where the MiG-17F scored kills of 100% against Navy fighter pilots. This shocking revelation occurred on their first MiG-17 challenge during the exploitation at Groom Lake. On their first encounter, the Air Force did not score much better.

These embarrassing encounters resulted in the Navy creating the TOP GUN Weapons School in 1969, which produced strong results against the MiG-21 when they encountered it in 1972.

The Air Force followed the Navy's lead by creating a dissimilar air combat program in 1972/73 when it created RED FLAG to give its pilots a better edge in the fight.

A year earlier, both the USAF and USN had learned through the HAVE DOUGHNUT exploitation to avoid prolonged maneuvering engagements, aka dogfighting. HAVE DOUGHNUT encouraged emulating the MiG-21's hit-and-run tactics.

The MiG-21 encounters over the skies of Nevada developed dozens of ways in which American planes proved superior to the Soviet fighter. The F-4E Phantom proved this while flying against the MiG-17F in 26 sorties. The F-105D and F-105E Thunderchief did so while flying 18 sorties. Despite its sleek shape, the F-4, F-105D, and F-104 found the MiG-21's performance at high-altitude inferior. Exploitation of the MiG-21, which fought with missiles, did not solve the problem of the US continuing to lose the air war in Vietnam against the older MiG-17F that shot only cannons.

Technical Exploitation of the MiG-17F Fresco

As part of the technical phase of the exploitation, AFFTC personnel installed the instrumentation to obtain quantitative performance and stability data of the HAVE DRILL MiG-17. Along with the airborne data acquisition system, they installed various temperature and pressure probes to measure engine inlet conditions during engine ground runs.

Installation of the test equipment in the MiG-17F began on 28 January 1969 and completed on 16 February 1969; with approximately 960 man-hours required for the installation. This figure did not reflect the in-shop time expended by the AFFTC calibrating test instrumentation components, fabricating special adaptors, fittings, fuel lines, etc., which could not be accomplished onsite due to the lack of the specialized equipment required to accomplish these tasks.

Installation and Alteration

- A UHF antenna, wiring, mounting, brackets, and radio set installed.
- A tape recorder installed in the cockpit.
- Cockpit installations and modifications included the "G" meter on left-hand aux. panel.
- Intervalometer on right-hand aux. panel.
- Airspeed Indicator-Main panel.
- Altimeter-Main panel.
- Machmeter in place of compassed radio indicator.
- Removed ships EGT and Tach indicator, calibrated and reinstalled.
- Removed V.H.P. Control picked up plugs and wiring to go through the cockpit pressure area to the nose section for power leads.
- Installed instrumentation circuit protector panel,
-
- Nose section installations and modifications include the following:
- Removed VHF radio and installed a photo panel.
- Pitot-static system mounted on the nose and spliced in at wing disconnect.
- Outside air temperature probe mounted on oxygen servicing door.
- X-band antenna mounted on oxygen servicing door.
- Gun platform installations and modifications include the following:
- Ammunition cans removed.
- Manufactured brackets for and installed a Midwestern oscillograph model #581 fourteen channel.
- Signal conditioner package mounted near oscillograph.
- A K-3 Minneapolis Honeywell attitude gyro installed.
- Three rate gyros installed.
- Two Potter flowmeter amplifiers and totalizers installed.
- Oscillograph counter timer delay box installed.
- Fuel amplifier for oscillograph mode installed.
- X-band beacon and power divider installed.
- X-band antenna installed.
- Accelerometer transducer installed.

TEST EQUIPMENT WITH MIG-17 AIRCRAFT

Engine Compartment Installations and modification Included the Following.
- Flowmeter transducer installed in the main line.
- Flowmeter transducer installed in afterburner line.
- Wiring tapped into the ship's tach generator.
- Control Surface Position Installations.
- Market potentiometers and brackets mounted on the elevator torque arm,
- Market potentiometers and bracket fitted to the rudder torque arm.
- Markite potentiometers and bracket mounted on left-hand aileron torque arm.
- Technical exploitation objectives

The objective of their joint exploitation efforts focused on data acquisition internal to the plane and by the special projects exploitation team tracking the planes during the tactical phase to follow. The data collection included manual, voice recording, photo recording, and electronic data by way of an oscillograph. In the MiG hangar, AFFTC installed an external observed r's panel to record engine ground run data in addition to photographing everything by a hand-held 35 mm camera.

AFFTC installed manual system-calibrated test instruments to replace the standard aircraft instruments on the pilot's panel. These instruments included:
- Clock: 8-day with a start and stop sweep second hand (Type A-13).
- Indicator, Airspeed: 50 to 850 knots, Kollsman Instrument Corp. (Type 739DX.)
- Altimeter: 0–80,000 feet, Kollsman Instrument Corp., Type AN5760–5.
- Machmeter: 0.5 to 1.5 Mach No., Kollsman Instrument Corp., Type A-2.
- Accelerometer: -4 to 10g, Bendix Aviation Corp., Type B-6.

In addition to the above, AFFTC calibrated the standard aircraft engine RPM and exhaust gas temperature indicators, calibrating the EGT as a system using a Jet calibration tester. The AFFTC personnel obtained the proper heater probes by sending a sample thermocouple probe to AFFTC at Edwards to check against various probes on hand.

The exploitation team recorded pilot comments by installing a Wollensak Model 4200, cartridge type magnetic tape recorder on the aft end of the right-hand vertical console. AFFTC mounted this recorder on a quick-release mount to facilitate changing tape cartridges and to provide access to the various ground servicing valves. The input to the recorder was through a special adaptor installed between the pilot's helmet disconnect and the aircraft's disconnect cable. Use of this adaptor allowed the pilot to start and stop the recorder as required and enabled recording of all transmissions and receptions when the recorder was running.

Along with the Wollensak, Model 4200 recorder installed in the aircraft, AFFTC test engineers used another Model 4200 in the Control Tower as a backup measure to record all transmissions and receptions during USAF test flights.

To record basic performance data, AFFTC installed a 12-parameter folded photo recorder laterally in the nose bay between fuselage frames 3 and 4in a space made available by the removal of the aircraft's standard ADF, VHF, and IFF radio transceivers.

AFFTC used a Traid Corp., Model 1000B 16 mm pulse-recording camera equipped with an Angenieux, 10 mm, fl.8 lens. They used an LB4BT, 50-foot capacity film magazine, to photograph this panel, and Eastman Kodak Plus X, l6mm film as the recording media.

Twenty-two GE 1495 type 28 VDC bulbs mounted on the face of the panel provided indirect lighting of the test instruments. They painted the top 1/3 of each bulb with high-temperature silver paint to prevent glare into the camera lens and to act as a reflector.

The pilot selected a USAF B9A intervalometer mounted at the top right-hand side of the pilot's instrument panel. He selected it as required to pulse the photo recorder camera at a pulse rate selectable from two frames per second to one frame every 60 seconds.

AFFTC personnel mounted an oscillograph, and associated signal conditioning equipment on the gun pod between aircraft frames 4 and nine by removing the ammunition containers and installing special provisions for mounting the test equipment. An oscillograph run switch mounted on the test instrumentation control panel controlled the oscillograph. Along with this switch, AFFTC mounted an OSC FAIL indicator light next to the OSC RUN switch. This light illuminated to indicate an oscillograph malfunction or exhaustion of the oscillograph paper. The oscillograph system recording the stability parameters while a Midwest Instruments, Inc., Model 581, l4-channel oscillograph recorded the fuel flow data.

AFFTC measured engine, and afterburner fuel flowed using Potter Aeronautical Corp. turbine type fuel flow transmitters, Model 130 converters, and a special AFFTC designed oscillograph signal conditioning module.

Modified Model 130 converters converted the cyclic output of the fuel flow transmitters to suitable outputs for oscillograph and photo recorder presentation. Sodeco 5-digit counters presented fuel used data on the photo recorder with each count representing approximately 1/10 gallons of fuel used, presenting the fuel flow rate data on the oscillograph as a cyclic function. To aid in data reduction, a pedestal imposed on the oscillograph every 64 cycles, with each 64 cycles representing approximately 1/10 gallons of fuel used. Using the oscillograph timelines, AFFTC determined fuel flow by referring to the fuel flow calibration chart plotted as CPS against flow rate.

AFFTC also fabricated new fuel lines incorporating Potter flowmeters prevented modification of the aircraft's standard fuel lines and enabled adapting from metric size to standard AN size tubing and fittings. A one-inch diameter transmitter installed in the main fuel supply line between the tank outlet and the fuel filter inlet measured total fuel supplied to the engine and afterburner. They measured afterburner fuel flow using a 3/4-inch diameter transmitter. They mounted it in the afterburner fuel line between the afterburner fuel control unit and the afterburner spray ring inlet. It obtained the afterburner operating, engine fuel flow by subtracting afterburner fuel flow from the total fuel flow. Fuel evaluations included fuel temperature and outside air temperature.

To evaluate engine RPM, AFFTC removed the standard aircraft RPM Indicator and tachometer generator from the aircraft and calibrated them as a system at the AFFTC facility at Edwards AFB along with calibrating the standard aircraft tachometer indicator,

AFFTC displayed the parameters of indicated airspeed and altitude in the photo recorder on Kollsman IAS and altimeter indicators. These instruments connected using flexible surgical tubing and tee fittings to the aircraft's normal wing boom total and static pressure systems.

To record control surface positions, AFFTC recorded elevator, rudder, and left-hand aileron positions recorded on the oscillograph, measuring them using Marklte 5000 ohm, 357-degree travel potentiometers as transducers. They measured rudder and elevator positions using a cable and pulley system tied directly to the control surface torque rod. The left-hand aileron position used a cable and pulley system tied to the control linkage just inboard of the aileron.

AFFTC also used the oscillograph to record aircraft bank and pitch angles, measuring them using a Honeywell K-3 cageable attitude gyro, type JF7044A-35 mounted on the gun pod. A gyro cage-uncage light and switch mounted on the test instrumentation control panel provided data on the gyro limits.

The oscillograph recorded roll, pitch and yaw rates as measured using Humphrey rate gyros. The oscillograph recorded aircraft normal acceleration recorded using a Statham A43–10–350 unbonded strain gage transducer. The transducer mounted on the aft bulkhead of the gun bay and marked an event condition with a special push-button type switch mounted on the front left-hand top side of the pilot's control stick. Actuation of the switch deflected a trace on the oscillograph and illuminated a light in the photo recorder.

AFFTC used a 14 channel, modular type signal conditioning package for conditioning of the various transducer outputs to acceptable levels for oscillograph recording. The package provided automatic in-flight "R" calibration capabilities, also. AFFTC used a signal conditioner designed and built by for use in the Test Pilots School stability seat packages.

AFFTC conducted four engine runs to obtain quantitative, temperature and pressure information of various engine parameters, and to determine installed engine thrust. Two engine ground runs obtained engine thrust and static pressure, temperature information-one to get dynamic measurements of pressures and temperatures for the Navy Propulsion Lab and one to get fuel flow data at various power settings. Making necessary the last run was the failure of the oscillograph to record during the engine thrust run, with resultant loss of fuel flow data.

The exploitation team used two different systems of tie-down to secure the plane. The first method used two, approximately 20-foot long 3/8-inch diameter, steel cables. The cables attached to the main gear up-lock fittings on one end. The other ends shackled to another cable approximately 20-feet long running to the tie-down attachment point.

This system resulted in the cables attached to the main gear up-lock fittings riding on the main gear tires, placing the strain link used to measure thrust directly in the exhaust blast.

To eliminate these problems, AFFTC fabricated new cables and fittings to secure the plane for thrust measurements using AFFTC designed and fabricated 15,000-pound strain links inserted in the tie-down line. They connected Mayberry Instruments, Inc., multiple turn self-balancing indicators, calibrated in pounds of thrust (0–7,000 lbs.) to the strain links through a 30-foot electrical lead. They mounted the indicators at the external observed r's station and manually recorded the readings. This method appeared as a repeatable and accurate method of obtaining thrust measurement in the field. The testing personnel reduced hysteresis error at each stabilized point by physically rocking the aircraft forward and aft at the left-hand wing tip.

The Naval Air Propulsion Test Center required the measuring of various engine parameters during throttle bursts for which the exploitation team conducted one engine ground run. NAPTC paralleled its test recording equipment into existing test instrumentation systems to measure compressor inlet pressure, engine RPM, and fuel flows. The Navy installed an X-Y plotter and a CSC direct-write oscillograph in a radio-equipped truck to enable communication with the aircraft operator during the engine run.

The Navy combined this run with the thrust measurement run to conserve engine ground run time. FTD personnel manually and photographically recorded the tests using an open face test instrumentation panel mounted in a radio-equipped vehicle. The panel interconnected to the aircraft through 40-foot long pressure tubes and thermocouple leads.

Along with the pressures and temperatures measured on the engine, FTD personnel fabricated a pressure, temperature rake containing three pressure probes and one temperature probe to measure exhaust gas pressure and temperature at the exhaust nozzle exit. Mounting the rake on a type B2 bomb hoist enabled the removal of the probe from the gas stream during afterburner operation. The outputs of this rake connected to the test engineer's panel through 40-foot pressure and thermocouple leads.

The Aeronautical Systems Division and Research & Technology Division conducted the systems evaluation, vulnerability study of the MiG-17F as an exploitation team effort. The effort included the combined results of exploitation conducted during project HAVE DRILL by:

- The Aeronautical Systems Division (ASD),
- Air Force Astronautics Laboratory (AFAL),
- Air Force Flight Dynamics Laboratory (AFFDL),
- Air Force Material Command (AFML),
- Air Force Aero Propulsion Laboratory (AFAPL)

ASD Laboratory objectives under this program included determining the technological state of the aircraft and its subsystems. The ASD Laboratory determined the vulnerability of the plane to the US current gun and missile threat, and determined characteristics of the installed radar's performance and its susceptibility to ECM.

The team found the hydraulic system design not readily vulnerable when subjected to conventional gunfire damage. The pilot could still fly the aircraft to safety with the systems completely disabled; however, if ignited, the flammable hydraulic fluid used in this aircraft proved capable of creating extensive aircraft damage.

The Soviet designers located most of the fluid in the system forward of the hot sections of the engine. Leaking fluid could readily ignite by the cooling air carried it into the hot part of the engine and be ingested it into the engine compressor. However, this area also contained the engine fire extinguishing system to abate any hydraulic fire and associated fire damage.

The plane stored air in both the landing gear strut for an emergency extension of the gear and emergency wheel brake operation and the armament service tank for gun charging and wheel brakes. The air stored in the strut and armament tanks rated at 710 psi with a volume of 402 and 122 cubic inches, respectively.

Regarding the difficulties met with, the exploitation team consumed approximately 75 percent of their time just correcting malfunctions. These were those necessary to maintain the radar at the level required to provide expected radar performance while the exploitation team has the radar available for exploitation.

The following is a list of the major test equipment used in the tests:

- 1 TS-1011UPM-84 spectrum analyzer
- 1 EH-121 pulse generator
- 3 EH-131 pulse generator
- 1 RF Generator
- 1 Polarad S-band, RF generator-HP 6l6A
- 1 Telonic signal generator (nautical miles 2000 with an SH-1 plug-in head)
- 1 Tecktronic 555 oscilloscope
- 1 Teckronic 535 oscilloscope
- 1 Hewlett Packard power meter
- 1 Simpson Model 260 multimeter
- 1 Hewlett Packard vacuum tube voltmeter

MiG-17 Performance, Stability, and Systems Evaluation

The AFFTC and NATC customers, in conjunction with the special projects exploitation team, conducted the performance, stability, and systems evaluation of the MiG-17F at Groom Lake.

Following the technical exploitation, the HAVE DRILL MiG-17F made its first US flight at Groom Lake in January 1969. The second MiG-17F code-named HAVE FERRY flew in March of that year.

Flying both planes, the exploitation team evaluated the performance, handling qualities and systems evaluation of the MiG-17F exploitation programs led by AFFTC and NATC. The exploitation team flew the two aircraft for test a total of 20.7 hours in 33 flights. The team equipped only the #1 aircraft with test instrumentation for obtaining quantitative data. Therefore, all data presented herein refers to the #1 (HAVE DRILL) MiG-17F unless otherwise stated.

HAVE DRILL/HAVE FERRY AIRCRAFT

The exploitation team, consisting of the CIA, Air Force, Navy, and Special Project contractor personnel, obtained the quantitative data using a clean aircraft with no external stores, limited quantitative data, and with two external fuel tanks (106 gal/tank). The engine start gross weight with no external stores was approximately 11,700 pounds and with two external fuel tanks was approximately 13,200 pounds. No attempt was made to vary the center of gravity, which eliminated the testing of the handling qualities at extreme g.

The exploitation team found the MiG-17F a simple, sturdy, inexpensive, easily maintained aircraft with excellent maneuverability capabilities at 450 KIAS and below. Canopy visibility was excellent except out the front horizontally as well as forward and down.

The periscope mounted on the canopy was a simple, however, excellent aid to the pilot by providing rear hemisphere clearance for the pilot with minimum head movement. This became particularly important in the high g maneuvering engagement. The simplicity of the aircraft and its systems made it possible for US pilots to start flying the plane at maximum performance after only two checkout flights.

The MiG-17F aircraft was a single-seat, swept-wing, point defense interceptor powered by a VK-IF centrifugal-flow jet engine with afterburner. The VK-IF engine had an uninstalled sea level static thrust rating of 7,440 pounds at maximum afterburning and 5,720 pounds at military per the manufacturer's data. Without stability augmentation, the plane powered only the ailerons of the primary control surfaces. It carried its fuel in two internal fuselage tanks of 1250 and 16 0-liter capacity and, optionally, from two external wing tanks of 400 liters each.

The VK-lF engine used a single stage double-faced centrifugal compressor with combustors consisting of nine cans. The turbine was a single stage, axial type. The exhaust nozzle was a hydraulically operated two-position convergent nozzle with a maximum afterburning diameter of 24.38 inches and a non-afterburning diameter of 21.75 inches. Maximum afterburner and military engine speed were 11,560–20 RPM with an exhaust gas temperature of 718 degrees C. Engine speed for minimum afterburner was approximately 10,870 RPM. Changing engine RPM accomplished thrust modulation between the minimum and maximum afterburner.

Each aircraft carried one 37 mm gun, two 23 mm guns, and a gun camera. The 37-mm gun fired 40 rounds at 400 rounds/minute, and each 23-mm gunfire 80 rounds at 900 rounds/minute.

For the evaluations, the exploitation team instrumented the HAVE DRILL MiG-17F for performance and handling quality tests. They instrumented the HAVE FERRY MiG-17F only with several sensitive cockpit instruments.

The cockpit of the aircraft was an antiquated design impossible to enter or exit the cockpit with any degree of urgency. It required the pilot to step on the seat type parachute, an integral part of the seat, supporting himself with his hands on the canopy rails while threading his feet onto the rudder pedal stirrups. It required an average of two minutes to don the parachute harness and hook up the necessary personnel leads after entering the cockpit.

Pilot restraint mechanism consisted of a semi-conventional lap belt and shoulder harness arrangement. The pilot found it difficult to clinch the lap belt tight due to an awkward out and downward movement of the tightening straps. The shoulder harness pulled tight in a conventional manner. However, it failed to stay tightly adjusted under normal pilot movement. The shoulder harness locked or released by movement of a locking knob located on the forward left side of the seat. The locking knob was opposite to USAF standards, moving forward to release the shoulder harness, while movement aft locked it.

The personnel leads (oxygen/communications, etc.) remained an integral part of the seat. Seat height was not adjustable and required padding for various sitting positions and percentile persons. The rudder pedals are adjusted manually before entering the cockpit. However, the pilot could adjust them after entering the cockpit. Pilots over 6 feet tall and husky in stature found the cockpit, the approximate size of an F-86F with less leg room, small, cramped, and difficult comfortably to fly the aircraft

Foreign Technology Division Cockpit Evaluation

FTD evaluation of areas that limited or enhanced cockpit visibility as viewed from the pilot's seat of the MiG-17F occurred through ground measurements of the cockpit visibility. These happened in a hanger using fixed reference points to measure and then calculate all angles.

A pilot sat in the cockpit simulating flight conditions by wearing a helmet, oxygen mask, and parachute with the pilot's safety belt and shoulder harness securely fastened.

The MiG-17F canopy provided good visibility above the horizontal plane. The gunsight restricted forward flight visibility through the windscreen and required pilot head movement to circumvent this obstruction. The canopy framework, antenna strips, and mounting assembly of the aft looking periscope slightly restricted upward, and the ejection seat headrest restricted visibility rearward.

The low pilot seating position and bulbous nature of the fuselage limited side visibility below the horizontal plane limited to approximately 40 degrees. The large size and position of the wings further degraded visibility to the aft and low positions.

The canopy-mounted periscope was an excellent aid to the pilot providing rear hemisphere clearance with minimum head movement. The canopy limited side-to-side head movement and the headrest limited aft head movement.

During climb performance checks, the exploitation team used both maximum afterburner and non-afterburner thrust with no external stores. The exploitation team performed the maximum afterburner thrust climbs from approximately 10,000 feet to near combat ceiling (500 feet per minute rate of climb of the aircraft).

Maximum afterburner thrust climb showed a combat ceiling of 51,560 feet, which agreed with the predicted data. At 36,000 feet, the exploitation team found the test day rate of climb for maximum afterburner thrust twenty percent less than expected.

Maximum non-afterburner thrust climb showed a less rate of climb than the predicted data. At 36,000 feet, the exploitation team found the test day rate of climb for maximum non-afterburner thrust approximately twenty-two percent less than anticipated.

The exploitation team found the engine thrust adequate for this aircraft and satisfactory engine response above 7000 RPM while making afterburner climbs to 51,000 feet without a blowout. The exploitation team used afterburner without problems up to 45,000 feet, except during one flight during which required the pilot to descend to 38,000 feet to obtain an afterburner relight.

The aircraft demonstrated outstanding maneuvering performance, both in maximum attainable g and high g level reached before buffet onset.

MiG-17 Combat flights at Area 51

The USAF Tactical Fighter Weapons Center conducted the tactical evaluation under the overall management of the FTD. The TFWC evaluated both the MiG-21 and the MiG-17 as total weapons systems and operating in a tactical environment to compare them with a variety of TAC aircraft the Air Force expects to confront a threat.

The US Navy likewise conducted a tactical evaluation to determine the capabilities and limitations of the Fresco C in the air combat maneuvering environment against Navy combat aircraft. The Navy's goal defined the area of comparative strength and weakness, and to determine the tactics necessary for Navy combat aircraft to defeat the Soviet MiG-17 Fresco C in the air combat maneuvering environment.

A year later, and following the MiG-21F technical exploitation, follow-on exploitation of the MiG-17F Fresco began under the code names HAVE DRILL and HAVE FERRY.

The Aerospace Defense Command and US Navy attack and fighter squadrons under the overall management of the FTD conducted both technical and tactical evaluations of the Fresco C. The evaluations occurred under simulated combat conditions against the F-4, F-5, F-8, F-100, F-102, F-104, F-105, F-106, A-4, A-6, and A-7 aircraft.

The AFFTC and NATC evaluated the systems of the two MiG-17F planes for performance, stability, and control. The Aeronautical System Division and Research & Technology Division Laboratories evaluated the MiG-17 systems and vulnerability. The Armament Development and Test Center, NWC, Naval Missile Center and FTD conducted infrared measurements. CIA contractor EG&G special projects conducted radar cross-section measurements and provided radar tracking of the MiG-17F for the FTD and Naval Missile Center.

To evaluate combat flights, the exploitation team made the cruise out and back at 20,000 feet at speed for best range. The exploitation team used their test data from the technical phase of exploitation and calculated two typical combat flights, one calculated for the clean aircraft and the other for two external fuel tank loadings.

The exploitation team based each flight on a three-minute sea level combat at afterburner thrust and a five-minute military thrust sea level combat. The exploitation team allowed five minutes for warm up and taxi which used 70 liters of fuel. Takeoff and acceleration took one minute and used 62 liters.

The exploitation team used the same intercept and returned headings for all the climbs and descents. The exploitation team expended 268 rounds of ammunition during the tests. Descent to combat took three minutes and used 53 liters. The exploitation team allowed five percent reserved fuel, 71 liters for the clean aircraft and 110 liters for external tanks, which the pilot dropped when empty.

Combat radius of the clean planes and three-minute afterburner thrust sea level combat covered 115 nautical miles, while the combat radius for the external tank loading covered 215 nautical miles. With external tanks, the combat radius extended 100 nautical miles, approximately 87 percent. The time for intercept for these two flights was 18.6 minutes for the clean aircraft flight, while the external tank loading flight required 36.2 minutes.

In summary, although the heavy control forces above .85 Mach and the high-altitude pitch up tendency just described as tactical limitations, the MiG-17F possessed turn capability significantly superior to any US jet fighter aircraft.

The vast majority of tactical engagements against the MiG-17F in SEA occurred in the low altitude regime with the Fresco C low wing loading and 8.0 g structural limit best optimized and used.

The plane's outstanding maneuverability in this area permitted this rather old and simple fighter aircraft to remain such a potent threat in the day of sophisticated modern weaponry.

The evaluation team evaluated rolling performance by measuring lateral control with aileron boost ON over a wide range of Mach numbers. These occurred at 10,000, 20,000, 30,000, and 40,000 feet with the aircraft in the clean configuration.

The exploitation team conducted limited testing with the aircraft loaded with two external tanks, with aileron boost OFF and with partial lateral stick deflections. The clean aircraft with aileron boost ON exhibited poor rolling performance of peak roll rates only between 100 and 130 degrees/second. These peak roll rates occurred at .60 Mach at 10,000 feet, and .65 Mach at 20,000 feet, and 80 Mach at 30,000.

Peak roll rates did not vary appreciably with altitude. However, they did decrease rapidly at airspeeds both below and above the Mach number for the peak roll rate. Adverse yaw was practically nonexistent at all conditions tested.

The addition of two empty external fuel tanks to the aircraft did not affect the peak roll rate. However, the initial roll response was slower as witnessed by increased time to bank 90 degrees and decreased bank angle in the first second when compared to clean aircraft data. The exploitation team did not conduct lateral control tests with fuel in the external tanks.

The primary objectives of the test program determined the tactical capabilities of the aircraft and accomplished a limited performance and stability evaluation. The exploitation team did not conduct specific system tests, relying instead on observations and a qualitative evaluation of system performance as follows:

The gun charging system permitted inflight gun loading by the pilot to reduce reaction time required to launch an interceptor force. The design eliminated the need for ground gun-arming procedures and safety, allowing the pilot to clear gun jams in flight.

Several times during the test program, the gun jammed during firing passes. Actuation of the gun charging system cleared the jam and permitted further gun firing.

During these firing passes, the pilot did not notice any excessive pipper jitter, excessive cockpit noise or gun gasses in the cockpit. The only problem area was the bright flashes from the gun muzzle blast during firing as disconcerting to the pilot during night gunnery flights.

The exploitation team found the armament and electronic systems designed primarily for the limited day fighter role. The SCAN FIX radar only provided range information (a green light gave an in-range indication) and was effective to one nautical mile. The system required the pilot to use the optical sight for azimuth information.

One problem area encountered was determining which target the radar locked on in a multi-target environment. It required the pilot to break the lock, reacquire the target, and then determine if the new target and the range cue on the sight coincided. The pilot lost valuable time in a fast-moving aerial combat maneuvering situation while performing this task.

The armament consisted of one 37 mm cannon and two 23 mm cannons mounted on a lowerable carriage beneath the cockpit. The gun carriage came equipped with a hoist, which consisted of a winch, four cables, and a system of rollers. Ground personnel used a special wrench to operate the winch system for loading and maintenance.

The exploitation team found this a convenient method to gain access to a system usually well submerged in an aircraft's fuselage. However, the exploitation team found it difficult to ground clear the guns following a firing mission. A shell casing remained in the chamber of each gun and took an excessively long time to remove. (This difficulty most likely resulted from not having the proper armament tools to perform this operation.) The cannons rigidly mounted at two points to prevent recoil. A third supporting attachment fastened the barrels to the fuselage bulkhead by clamps to reduce dispersion.

The reliability of the aircraft rated outstanding. The HAVE DRILL vehicle accumulated 131–3 hours during 172 flights in 87 elapsed days. The exploitation team flew four to five flights daily and could have flown more except had the exploitation team not limited the flights to daylight hours and required briefing and debriefing time between flights.

Never during the evaluation period was the aircraft flown as often as possible during an operational maximum-effort flying schedule. The exploitation team lost some flight days or flyable missions because of bad weather and practically nonexistent logistics for a black program and a foreign-built plane. Considering the secrecy conditions under which the exploitation team conducted the evaluation, the reliability record of the test aircraft was not only exceptional, however, was a sobering fact.

A weight and longitudinal balance operation conducted before the first flight ensured a safe and representative aircraft. The exploitation team obtained the desired center of gravity limits from the airframe manufacturer's operational curve of gross weight versus, e.g., for the MiG-17F.

In conclusion concerning of their evaluations, the exploitation team found the MiG-17F demonstrated outstanding lift-limited maneuvering capabilities with high available g and g-level for onset buffet and a low turn radius.

Its internal fuel capacity, with no external stores, limited its range and endurance to a 115-nautical mile combat intercept radius, including a five-minute warm-up and taxi. Following this, a military climb on heading to 20,000 feet and cruise at that altitude, followed by three minutes of combat in afterburner near sea level. The aircraft took a return heading military climb to 20,000 feet, made an idle descent with 5 percent fuel reserved. With wing tanks, the radius extended to 215 nautical miles with the wing tanks dropped after depletion.

The exploitation team found the aircraft very reliable with exceptional operational availability. The exploitation team easily accomplished four to five flights day after day with an operational record resulting from a deliberate design approach toward airframe, engine and system simplicity and reliability. Attractive features included dependable in-flight gun clearing and charging, a rear-view periscope, and a smooth, well-balanced flight control system. Particularly impressive was the lack of engine exhaust smoke at lower altitudes.

The MiG-17F was easy to fly, and the undesirable features, such as spin and accelerated stall, readily handled when encountered. With proper user knowledge of its maneuvering capabilities and limitations, the MiG-17F was a very effective interceptor/air superiority daylight fighter throughout most of the subsonic flight envelope.

Radar cross-section Evaluations

The CIA's EG&G special projects contractor exploitation team performed the MiG-17F radar cross-section measurements for the USAF FTD and the Naval Missile Center, Point Mugu, California as part of the technical exploitation. For this phase of the exploitation, the radar cross-section measurement system composed of six basic radar subsystems controlled by one reference bull-gear servo assembly designed by EG&G.

The EG&G Special Project Exploitation team operated and monitored the measurement evaluations from a master control facility in a converted two-story former Navy barracks. The facility, located next to the RATSCAT radar array, used only four of the radar systems in the measurement program.

The exploitation team used the VHF, S, and C-Band equipment to gather the reflectivity data, and the Nike X-Band system as the target tracker to generate range information and target bearings. The special projects team designed, constructed, and programmed a special computer to interface all systems to ensure data acquisition and performance integrity.

The exploitation team slaved the other radars to a reference servo network controlled by target bearing data from the X-Band radar. The exploitation team digitized the range-gated video data received by each radar system and recorded it on tape for offline computer processing. The exploitation team did not use their VHF telemetry system that normally provided target pitch, heading, and roll data.

For the Navy, the exploitation team on the special projects Exploitation team conducted X-band cross-section measurements using their CIA Nike X-band radar. The Navy Weapons Center especially to control the flights and obtain the nose and tail data during radial passes followed by offset passes to get broadside data. The plane conducted combat maneuvers to obtain several other aspects. The special projects Exploitation team obtained data on several other flights on a noninterference basis, obtaining aspect angle information on these flights by a boresight camera.

Following the cross-section measurements for the Navy, the exploitation team conducted six flights for the US Air Force FTD to obtain VHF, S-band, and C-band cross-section data. During three of the flights, the exploitation team obtained broadside data. On three, the exploitation team obtained nose and tail data, recording the pulse-to-pulse recordings on digital tape where the exploitation team reduced the digital data to yield median values.

The Special Project exploitation team officially conducted the MiG-17F airborne flights per the specifications and perimeters established by their Navy and Air Force customers. However, for the exploitation team, this allowed them the opportunity to advance RCS technology from that obtained during Mach-3 overhead flights of the CIA's A-12 Project OXCART stealth reconnaissance plane. They also conducted RCS evaluations of the MiG-21 Fishbed. At this point, the special projects Exploitation team knew that another RCS program loomed on the horizon — the Project HAVE BLUE stealth prototype that became the F-117 Nighthawk.

Besides the M-33, mobile Nike X-band fire-control radar, the Area 51 special projects data acquisition system used essential recording equipment that included a recording oscillograph and an FM tape recorder that recorded target range, AGC voltage, pulse amplitude, azimuth angle, elevation angle, and timing.

The transmitted signal, coupled to the antenna through a duplexer where a test coupler monitored it while injecting test signals from an X-band signal generator.

A 60 MHz intermediate frequency (IF) amplifier controlled by an AGC-system amplified the received signal with an amplitude modulation 10 Hz upper cutoff frequency. The output of the receiver fed into a boxcar detector, a sample-hold device with a return to zero, effectively reducing the bandwidth to 1.5 kHz and amplitude for recording on the oscillograph and tape recorders. The output of the receiver drove the angle and range-tracking systems. A digital-to-analog (D-A) converter converted the digital range information from the range tracker with the linear voltage varying between 0 and 4.5 volts proportional to the target range.

The exploitation team received a 2048-yard ambiguity of the range readout resolved by the exploitation team observing a coarse analog range readout obtained from the analog computer as a dc voltage proportional to range. The exploitation team recorded the range output signals on the oscillograph and tape recorders along with the analog computer's height, ground range, heading and provided inputs for an X, Y, Z plotting board, all synchronized by a timing pulse generator.

For visual observation, optical tracking, and photography, the exploitation team bore sighted an optical system, this consisting of three periscopes with the radar antenna providing a field-of-view of 6 degrees and a magnification power of eight.

For a dynamic cross-section of FTD audio-frequency measurements collected against the HAVE DRILL vehicle, the exploitation team used two NAGRA III tape recorders and three sensing devices for collection.

- A dynamic microphone for recording sounds in the usual manner,
- An infrared radiometer recorded the audio frequency modulations of the infrared radiation from moving targets, and
- A CW radar recorded modulations of the radar return from the target

Chapter 8 - **AFFTC and USN MiG-17F Propulsion Systems Evaluation**

AFFTC Propulsion Systems Evaluation Performed by FTD Naval Air Propulsion Test Center.

For the AFFTC MiG-17F Fresco C propulsion system evaluation, the exploitation team chose the HAVE DRILL plane serial number 055 to obtain data from the engine test and collect engine installation information. The VK-1F power plant for the Fresco C aircraft was a 1965 Poland-built afterburning turbojet derivative of the British Rolls-Royce Nene engine and its US version, the J42-P series engine. The J48-P series engine was an improved and uprated development of the basic Nene power plant.

The VK-1F engine model specification indicated a guaranteed life for the engine of 100 hours. However, other information stated an increase to a standard of 200 hours, and in actual practice, 250 hours are frequently elapsing before engine replacement. The exploitation team believed the engine tested came equipped with original factory equipment. The exploitation team accomplished no engine trim before, or immediately after the HAVE DRILL test program proper.

Operational Test Exploitation of the MiG-17F

Lt Col Wendell H. Shawler, chief of the 6512th Test Squadron's special projects Branch at Edwards Air Force Base, served as project manager. VX-4 test pilots LCDR Foster S. "Tooter" Teague, and LCDR Ronald F. "Mugs" McKeown drew up a test plan. Fred Cuthill made the first functional check flight in the HAVE DRILL MiG-17F (YF-113A) on February 17, 1969, the first of 172 technical and tactical evaluations sorties that concluded in the middle of May.

The second MiG-17, code-named HAVE FERRY and designated YF-114C arrived on March 12, 1969, and completed its first functional check flight on April 9, 1969. This MiG, later, crashed while flying for the Red Eagles at the Tonopah Test Range, the crash killing Lt. Hugh Brown, US Navy.

The two MiG-17s flew against numerous US aircraft types during the tactical evaluations. In slightly over a month, the test exploitation team flew 52 sorties in the YF-114C, including dual flights with the YF-113.

The operational test exploitation program consisted of approximately 130 flying hours (from brake release on takeoff to five minutes after landing), 152 test flights, and some ground runs. By the end of the entire program, including flights after the test program the engine had acquired 209 flights and 155 total hours.

During the test program, the exploitation team made approximately 260 starts on the engine, encountering only three unsuccessful starts during the program, except for 20 to 30 attempts to get the starter to engage on 1 March 1969. The exploitation team attributed the 1 March starting problem to a faulty Microswitch while attributing a malfunctioning circuit breaker as the cause of the other three unsuccessful starts.

Discrepancies with the Engine Related Systems During the test:

On 21 March 1969, replaced the cracked and distorted afterburner tail cone with the afterburner assembly from another VK-1F aircraft A flow check of the removed afterburner fuel nozzles indicated low delivery in several of them.

On 17 April, the afterburner failed to light off at 17,000 feet, however, accomplished a successful light off at 10,000 feet.

During an engine ground test on 19 April 1969, the afterburner failed to light off on four successive occasions. The exploitation team canceled further attempts for light off and found the afterburner igniter misaligned, this most likely occurring due partially to overtemperature expansion during an earlier portion of the ground test.

On 22 April 1969, the maintainers removed, repaired, and installed a malfunctioning circuit breaker. At the time of disassembly, the maintainers found a hole in the aircraft aluminum liner near the afterburner light off zone caused by the liner melting during local overtemperature conditions during the 19 April ground run. Aluminum had splattered on the engine compressor inlet rear face.

On one occasion, the engine nozzle failed to open for afterburner operation. The exploitation team found that an electrical solenoid, mounted on an internal portion of the aft section of the fuselage, was not porting hydraulic fluid into the hydraulic pistons which connected by rods to the nozzle ring.

The exploitation team conducted two engine ground runs during the test. On the first test, the exploitation team recorded steady state values of compressor inlet pressure and temperature, compressor discharge pressure, turbine discharge pressure and temperature, and exhaust nozzle temperature and pressure.

The exploitation team recorded the fuel flow on an oscillograph, RPM and fuel temperature on a photo panel and a thrust indicating device reading thrust measured by load cells. The second ground test runs divided into two parts. The Air Force conducted a steady state engine test, and the Navy conducted a transient engine test aided by the Air Force.

Data acquired during the second steady-state ground test run was as outlined for the first ground run, with the addition of turbine discharge pressure.

The exploitation team presented only the data generated from the second steady stage engine test run, finding no apparent significant deviation of engine performance during the two tests.

The data taken during the second test also included thrust, a parameter that had anomalies in measurement on the first test. For comparison, the exploitation team tabulated only the data obtained along with published Soviet uninstalled data.

Based on the data acquired during the HAVE DRILL test, the exploitation team concluded the engine performance close to that of a new factory trimmed engine even though the engine had a high time. Soviet specification engine thrust and corrected engine thrust test values received during the test were in excellent agreement,

The nozzle pressure at military power was in exact agreement with the Soviet published data. Nozzle pressure was a parameter in which the exploitation team placed a high degree of credence as to proper engine trim and operation.

The VK-1F showed slow engine acceleration times in comparison to Free World operational engines of the same type and vintage. They remained, however, within the maximum times published in the Soviet literature. Engine reliability throughout the test was excellent; however, the pilots had trouble with engine controls.

Naval Air Propulsion Test Center VK-1F Engine Static Transient Ground Test

Consistent with the general objectives of Project HAVE DRILL, the Naval Air Systems Command assigned the Naval Air Propulsion Test Center (NAPTC) a task of evaluating transient characteristics of the VK-lF engine.

The Navy measured compressor discharge static pressure using a transducer tied into the fuel control sensing line. They measured the total fuel flow and afterburner fuel flow by turbine type flowmeters, which required AAFTC modifying the engine plumbing to facilitate these flowmeters. For the ground transient tests, AAFTC disconnected the recording equipment to record the fuel flow meter and engine speed signals using NAPTC equipment.

They recorded engine speed by measuring the frequency output of the engine tach generator.

To measure the turbine discharge pressure and temperature, NAPTC-fabricated four combination pressure-temperature probes to replace the standard exhaust gas temperature (EGT) probes. They connected the average output of three of the EGT thermocouples to an Air Force Howell temperature indicator installed in the cockpit. The remaining thermocouple output inputted to transient recording equipment, and one total pressure-integrating probe connected to a transducer.

AFFTC and the special projects exploitation team recorded all parameters on a 10-channel oscillograph, and total fuel flow and engine speed cross-plots on an x-y plotter in the special project's facility.

AFFTC flew three runs with the MiG-17F and found the VK-lF engine somewhat poorer than the J48-P series engine, both engines developed in approximately the same 1947 to 1952 period.

Preflight Phase (HAVE FERRY)

The HAVE FERRY aircraft arrived at the site on 12 March 1969. Assembly began on 20 March 1969 and completed in 20 days. During assembly, examined the aircraft for discrepancies followed by a 50-hour inspection where the team noted, repaired, adjusted, and operationally checked the following discrepancies:

- Right main tire worn beyond limits
- Right wing fence bent
- Leads on starting sequence control box loose
- Bottom nozzle actuator leaking
- A hole burned through the aft section shroud left side, 9 o'clock position and cracked.
- Aft section shroud cracked at 6 o'clock position
- Rivets loose both sides, bottom row, forward fuel cell (8) Left and right struts (Main gear) failed to hold air or hydraulic fluid.

During the reassembly and 50-hour inspection, inspected, adjusted repaired and operationally checked all systems. Also during this time, the exploitation team installed a UHF antenna, wiring, mounting brackets, radio set, and installed a tape recorder in the cockpit

Flight Phase (HAVE FERRY)

The exploitation team flew the functional check flight for the HAVE FERRY MiG-17F on 9 April 1969. The following was a summary of delayed discrepancies that generated during the project.

- IFF gear removed from the nose.
- IFF control box removed from the cockpit.
- Ballast installed in nose compartment. (78 pounds)
- Maintainability (HAVE FERRY)
- Man-hours expended for maintenance on the HAVE FERRY MiG-17
- 56.5 man-hours for discrepancies
- 32.5 Total Man-hours for maintenance 26,0 man-hours for post flights and preflights (2) 37
- 7 Total Flying hours
- 52 Total sorties
- 2.18 man-hours per flying hour
- 1.5 man-hours per sortie
- 30 maintenance discrepancies.
- 20 Flying days
- 2.6 Sorties per flying day

HAVE DRILL used this aircraft primarily as the backup aircraft for the HAVE DRILL; however, it flew 25 dual tactical flights.

Scope of the HAVE FERRY Evaluation

SCOPE OF THE TEST: The scope of the evaluation of the Fresco C included the following items against the F-4, F-5, F-100, and F-105:
- Determination of: Most suitable offensive maneuvers.
- Moat suitable defensive maneuvers.
- Most suitable flight tactics.
- F-4 radar detection data to include:
- Comparative-analysis of APQ-120 and APQ-100/I09 performance.
- Limited study of degree and 180-degree aspects.
- Altitude effects
- Verification and validation of:
- Tactical doctrine.
- Energy maneuverability comparative analysis.
- Operational restrictions and limitations of the MiG-17-
- Preparation of a briefing and motion picture adequately to cover the findings of the test.

MiG-17F Tactical Evaluation at Area 51

The HAVE DRILL and HAVE FERRY Maneuvers

The exploitation team flew twenty-six F-4 flights against the MiG-17F with the F-4. In most cases, they configured the F-4 with a full complement of air-to-air weapons to enhance the realism of the evaluation.

It conducted acceleration comparisons with one of three types: military power for both aircraft. Maximum power for both aircraft,
- Maximum power for the Fresco C while the F-4 maintained military power.
- They conducted both in level flight and in an unloaded condition.

The F-4 out-accelerated the Fresco C by roughly 50 knots, and 5,000 feet at military power, from 250 KIAS until the MiG-17F reached 400 KIAS. This represented a difference more pronounced at maximum power, approximating 100 knots, and some 8,000–10,000 feet.

When the Fresco C performed at maximum power, it displayed an acceleration potential somewhat better than the F-4 at military power.

Most the test runs occurred at 15,000 feet, although those that occurred at 30,000 feet did not expose any significant differences.

It was worth noting that on flight 1 and again on flight 49, the Fresco C initially responded better than the F-4, and the F-4 accelerated past approximately 270 KIAS to gain an advantage.

They performed zoom comparisons on flights 1, 2, and 19 to analyze the comparative zoom potential of the Fresco C. Both aircraft operated at maximum power, and with the Fresco C at maximum power versus military power in the F-4.

With both aircraft at maximum power, using the zoom profile resulted in an altitude advantage of some 4,000 feet and an airspeed advantage of 40 knots for the F-4 flight 3. When the Fresco C performed the maneuver at maximum power and the F-4 performed the maneuver at military power, the Fresco C enjoyed an 800-foot and 50-knot advantage.

Turn comparisons conducted on flights 1, 2, and 49; the exploitation team initiated comparisons at Mach .9, with two performed at 15,000 feet MSL and two at 30,000 feet MSL. It determined that at those airspeeds, the F-4 turning capability was very near that of the Fresco C if the Fresco C did not attempt to decelerate. The exploitation team also learned at a very early point that at airspeeds approximately 350

KIAS, the Fresco C had a turning capability far superior to that of the F-4. Consequently, the exploitation team expended no effort to ascertain that fact further.

The poor turning performance of the aircraft at high-calibrated airspeeds on previous flights precipitated the roll rate comparisons performed on flight 10. The comparisons occurred at 350 and 450 KIAS, 15,000 feet MSL. At 350 KIAS, the F-4 roll rate was twenty-five percent better (360 degrees - Vs. 270 degrees) than the Fresco C, and at 450 KIAS, the F-4 roll rate exceeded twice that of the Fresco C (360 degrees versus 160 degrees).

A single speed brake deceleration comparison occurred where the exploitation team initiated comparison at 3.5 KIAS from a stabilized line abreast position, 15,000 feet USL. The F-4 had 330 KIAS when the MiG-17F terminated the maneuver with 300 KIAS. The exploitation team concluded the Fresco C decelerated with speed brakes roughly twice as effective as the F-4.

The exploitation team attempted radar detection samplings on as many flights as possible, on the philosophy that a small sampling of radar data was inconclusive. Fifty-six recorded intercepts took place at various points throughout the evaluation. The exploitation team completed thirty intercepts using F-4C/D radars (APQ-100/109) and accomplished the remainder with F-4Es (APQ-120).

Within the definition of the data collected, the exploitation team found no significant difference between the performance of the APQ-100/109 and the APQ-120 in a clutter-free environment with approximately 22 nautical miles average at 25,000 feet.

This fact arose due to the ground clutter problem encountered by pulse radars, particularly when attempting intercepts over the mountainous or irregular terrain. The lower detection percentage of the F-4E could be attributable to the larger vertical beam width produced by the F-4E radar antenna.

A radar with a large beam width would, as could be borne out geometrically, encounter ground clutter problems at a higher altitude than a radar with a smaller beam width. At 10,000 feet, the F-4 detection probability 12.5% faltered badly. Insufficient parallel data existed for the F-4C/D to draw a similar conclusion. However, the trend indicated at 15,000 feet should have continued to decline at 10,000 feet.

Taken the average detection range for all passes on which detection occurred was 17.7 nautical miles. The significant point gleaned from this discussion was the performance of the F-4 radar at low altitude was unsatisfactory, and precisely where the Fresco C chooses to operate in Southeast Asia. RAW data collection occurred on Emissions 32–34. When the range only radar of the Fresco C operated, it always produced warning indications on the APR-25/26 or ER-14. The radar operated at a frequency of approximately 2,800 MHz, near that of the gun laying radar used by communist-bloc nations in Southeast Asia.

Therefore, on the APR-25/26, the AAA/AI light accompanied a strobe on the cathode ray tube. Accordingly, it became challenging to differentiate between AAA radar and the Fresco C range only radar in a multi-signal environment without an analysis of APR-25/26 strobe lengths. Such an analysis of a strobe in the rear hemisphere could disclose whether closure existed, by the eventual behavior of the length of the strobe, and could, thereby, serve to alert the crew of the possible presence of a Fresco C threat.

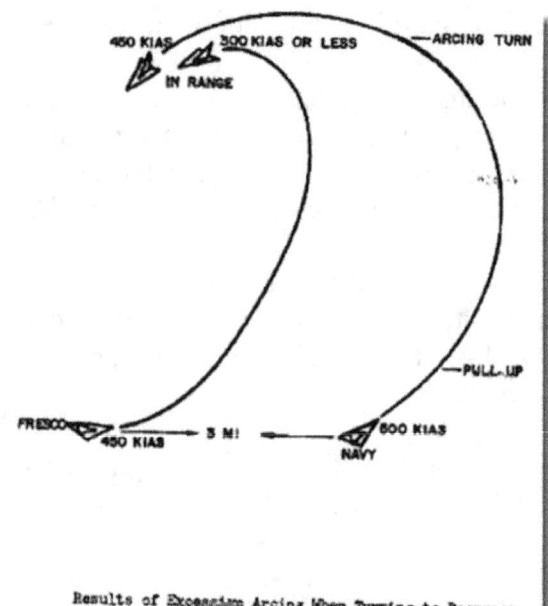

Results of Excessive Arcing When Turning to Reengage

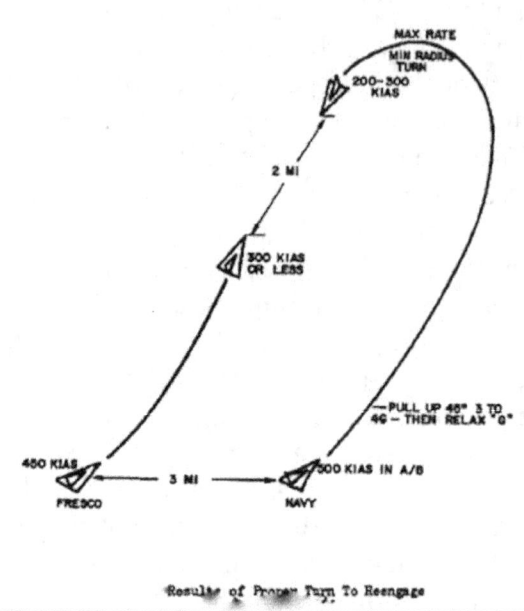

Results of Proper Turn To Reengage

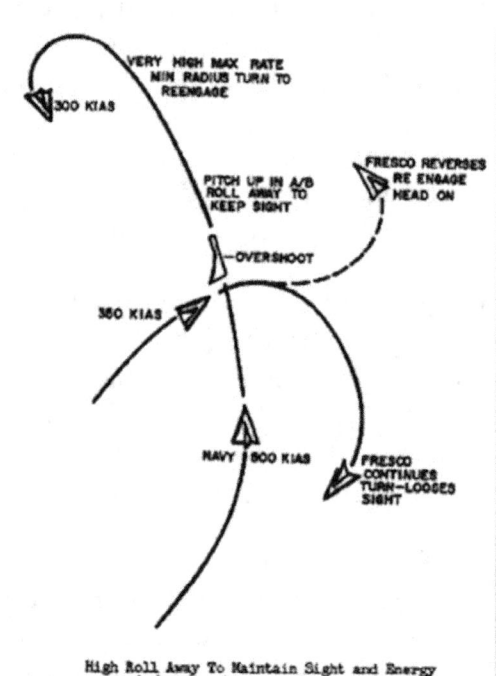

High Roll Away To Maintain Sight and Energy

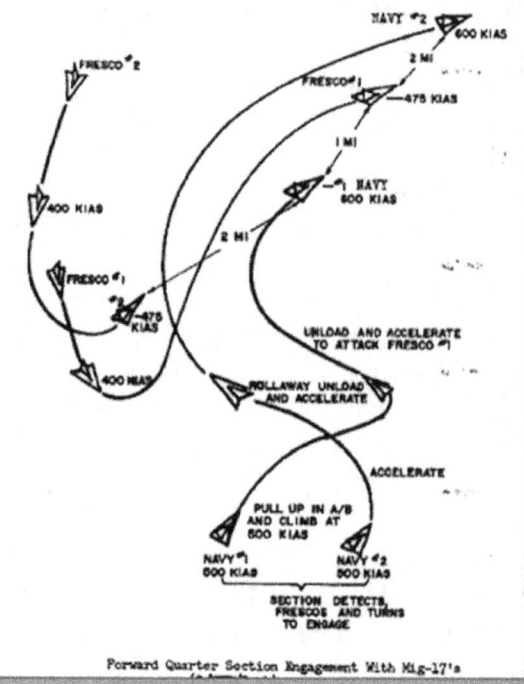

Forward Quarter Section Engagement With Mig-17's

Fresco C and F-4's

To summarize, the range only radar of the Fresco C illuminated the AM/AI light of the APR-25/26. Aircrews should carefully analyze such an indication before assuming an AAA radar lies at its source.

The exploitation team conducted offensive and defensive maneuvers by the F-4 against the Fresco C on flights 1–4, 10, 11, 16–21, 44, and 46–49. They thoroughly briefed these maneuvers and tightly controlled them to less than 360 degrees of turn. Offensive maneuvers included high/low-speed yo-yos, lag pursuit attacks, and maneuvers in the vertical plane. Defensive maneuvers included hard/break turns, slicing turns, climbing turns, level, and unloaded accelerations, high g barrel rolls over, unloaded rolls, and preplanned high-calibrated airspeed reversals.

In a one-on-one situation and with mutual awareness, the F-4 cannot turn with the Fresco C. An exception was when both aircraft operated at high-calibrated airspeeds as the maneuvers during flights 2, 3, 11, 19, 44, and 46 pointed out.

The exploitation team judiciously employed yo-yos and lag pursuit attacks because their continued application tended to permit the engagement to progress into low speed, high angle of attack maneuvering, very much for the Fresco.

Climbing turns proved marginally effective in permitting the F-4 to maintain an offensive potential. The MiG-17 demonstrated a better vertical capability than anticipated when if flew in afterburner power, flights 4, 10, 11, and 18.

Defensively, the F-4 could not generate a significant overshoot by the MiG-17 through hard/break turns into the attack. A slicing defensive turn was more effective than a level defensive turn insofar as it permitted the F-4 to retain energy while testing the ability of the Fresco C pilot more severely than a level turn. It was not a decisive maneuver.

The acceleration potential of the F-4 successfully produced a separation in defensive situation flights 3, 11, 17, and 18. It could have prevented any effective firing passed by the Fresco C if detection occurred early enough.

A lethal position forced the Fresco C to conduct some last-ditch maneuver to spoil its tracking solution before attempting an accelerated separation. Of such maneuvers attempted, the unloaded roll was the most effective in destroying a tracking solution.

A high g barrel roll produced the desired initial results. However, it diminished energy rapidly and left the F-4 behind the acceleration curve, unable to effect rapid separation. As an added precaution, once acceleration had produced airspeeds that forced the Fresco C above 450 KIAS, rapid reversals by the F-4 at random intervals denied the Fresco C lethal position as range opened.

Obviously, the F-4 advantaged from any escape maneuver that forced the Fresco C into a high calibrated airspeed situation. This inhibited the Fresco's maneuverability in that flight regime. Permitting unloaded rolls and unloaded accelerations to develop into steep dive angles while the Fresco C pilot concerned with the recovery of his aircraft beyond other considerations turned the advantage to the F-4.

To summarize, in a one-on-one situation, the F-4 crew should endeavor to arrive at an initial offensive position available. From that position, preferably in the rear hemisphere, weapons employment should observe the philosophy of increasing firepower with decreasing range, or missile attack followed with a gun attack if a gun was aboard.

If the F-4 failed to destroy the Fresco C, the F-4 should not become involved in a low airspeed, high angle of attack, turning battle with that aircraft. The F-4 should affect an immediate maximum power level or unloaded acceleration when gun-tracking was no longer possible.

The pilot could reattack if such a move were commensurate with flight objectives, by accelerating to Mach 1.2 or faster and executing a low (2–3) g vertical zoom while keeping the enemy in sight. Separation required to permit a turn-back should approximate three ITM. The retention of visual acquisition at that range required the pilot's utmost concentration. The pilot could expect subsequent attacks in the form of head-on passes unless the Fresco C pilot elected to fall off before the capability of his aircraft so dictates.

Defensively, outside the gun range of the Fresco C, maximum power acceleration could defeat a gun attack. Inside gun range, at 1,500 to 2,000 feet, the F-4 executed a last-ditch type of escape maneuver simultaneously with a maximum power acceleration to ensure survival. Such a maneuver should progress, if possible, into a steep dive angle accompanied by random reversals at high-calibrated airspeeds to deny the Fresco C a tracking capability during separation. Again, the pilot could reattack if commensurate with flight objectives.

During flights 16, 17, 19, 32, and 3, the exploitation team evaluated F-4 flight tactics employed against the Fresco C with the F-4 enjoying numerical superiority. Sound enemy tactics combined with competent Fresco C pilots presented a formidable problem to F-4 crews when the enemy had numerical parity or a numerical advantage.

The strategy employed against the Fresco C on these flights emphasized split plane maneuvering and mutual support between elements. Upon threat detection, the elements split in the vertical plane to force a decision on the MiG-17.

Once the enemy is committed, the superior acceleration of the F-4 came into play. The F-4s subsequently sandwiched the Fresco C while staying out of danger individually, and controlled the engagement until obtaining the desired result. The Fresco C could not disengage and separate under these circumstances without the F-4 firing upon it.

To summarize, the F-4 could defeat the Fresco C under conditions of numerical superiority by using split plane maneuvering to achieve a sandwich and complementing that tactic with mutual support. Mutual support was extremely critical. Crews not only kept constant pressure on the enemy, however, but also kept each other apprised of the enemy's position and actions. Elements could not separate where they cannot visually clear one another, especially in consideration of the fact that Fresco Cs rarely performed singly.

The pilot maintained a constant vigil in anticipation of the presence of a yet undetected threat. In all cases, the pilot employed high-calibrated airspeeds during tactical operations to complicate the attack problem for the Fresco C during all phases of flight when a Fresco C threat exists.

The following conclusions pertained to an F-4 aircraft configured with four semi-submerged inert AIM-7 missiles and four externally carried AIM-9 missiles and the Fresco C clean.

- With both aircraft operating at military power, the F-4 had an acceleration capability far superior to that of the Fresco C above 250 KIAS.
- With both aircraft operating at maximum power, the F-4 had an acceleration capability far superior to that of the Fresco C.
- With the Fresco C operating at maximum power and the F 4 operating at military power, the Fresco C acceleration capability was slightly superior to that of the F-4.
- With both aircraft operating at military power, the zoom capability of the F-4 was superior to that of the Fresco C.
- With both aircraft operating at maximum power, the zoom capability of the F-4 was superior to that of the Fresco C.
- The zoom capability of the Fresco C at maximum power was slightly superior to the zoom capability of the F-4 at military power.
- At 350 KIAS and below, the turning capability of the F-4 was inferior to that of the Fresco C, markedly so at airspeeds below 250 KIAS.
- At 450 KIAS and above, the F-4's turning capability compared to that of the Fresco C. However, the F-4 bled airspeed much more rapidly. If the Fresco C used speed brakes to decelerate while turning, it retained a turning capability superior to that of the F-4.
- With initial power settings closely matched and speed brakes employed, the Fresco C decelerated more rapidly than the F-4. An idle power deceleration without an accompanying utilization of speed brakes permitted the F-4 to decelerate at a rate comparable to the Fresco C.
- The F-4 roll rate with full-scale stick deflection was superior to that of the Fresco C, particularly at high-calibrated airspeeds.
- The F-4 could best defeat the Fresco C in an aerial engagement by using hit and separate maneuvers,
- If an attack was commensurate with flight objectives, the F-4 followed a separation maneuver with a high Mach, low g vertical zoom to regain an offensive potential,
- Defensively, the F-4 could readily defeat an attack by the Fresco C by accelerating to supersonic speed. If, however, the Fresco C was in such a position that an acceleration alone failed to produce safe separation immediately, the F-4 employed a last-ditch maneuver to defeat the lethal position of the Fresco C before attempting to separate. If possible, separation used a steep dive angle,
- Within the definition of the data collected on this evaluation, there was no significant difference between the Fresco C detection potential of the APQ-100/109 and APQ-120 radars in a clutter-free environment.
- All F-4 radars encountered ground clutter difficulties at altitudes of 10,000 feet AGL and lower. The Fresco C operating at or below 10,000 feet AGL substantially reduced the effectiveness of the F-4 air-to-air radar and radar missile system.
- The USAF and USN apprised F-4 tactical aircrews of the inadvisability of attempting a reattack in a one-on-one situation following an initial confrontation with the Fresco C and

a subsequent separation maneuver. The F-4 needed to obtain sufficient separation (2.5 to 3 miles) while maintaining visual contact with the threat. NOTE: If the aircrew lost visual contact, it required the utmost attention to reacquire the target visually. The crew resorted to radar acquisition.

- Keep F-4 ingress speeds sufficiently high to permit rapid acceleration to defeat a Fresco C rear hemisphere gun attack.

F-100 Versus MiG-17F Fresco C

Six test flights (8, 9, 15, 24, 25, and 53) evaluated the F-100 against the Fresco C, with only F-l00Bs in a clean configuration.

Acceleration comparisons performed found the F-100 accelerated above 300 KIAS before it achieved a degree of superiority in military power. At maximum power, the F-100 was clearly superior to the Fresco. If the Fresco C accelerates at maximum power and the F-100 accelerates at military power, the MiG-17 was slight, however, not significantly, superior.

Turn comparisons occurred on all F-100 flights, save number 15. The results of those comparisons varied with the power settings used and initial airspeeds. The F-100 turned faster than the Fresco C at 450 KIAS or higher. However, it bled airspeed more rapidly than did the Fresco C, which accelerated in a maximum performance turn at maximum power.

The two aircraft's airspeeds reduced to 350 KIAS as the two planes executed a 180-degree turn in a comparable period with the F-100 slightly better, as at higher airspeeds, though the Fresco C retained more airspeed. No turn comparisons occurred at slower speeds than 350 KIAS. However, it became apparent during the offensive and defensive maneuvers that a transition occurs between 350 KIAS and 300 KIAS, below the speed where the Fresco C enjoyed a dominant turn capability.

Roll rate comparisons occurring on flights 8 and 15 disclosed the F-100 roll rate at 350 KIAS as one-third better than the MiG-17. At 400 KIAS, the F-100 roll rate doubled that of the Fresco C. At 450 KIAS, the F-100 rolled level before the Fresco C could reach an inverted position (360 degrees of roll versus 160 degrees of roll).

Zoom comparisons performed on flights 15, and 2 demonstrated the F-100s slightly superior zoom potential at maximum power (30 knots and 500–800 feet). At military power, the two aircraft stayed comparable in a zoom, with a slight edge belonging to the F-100.

Comparative speed brake effectiveness analyzed on flight 8. Started from an initial 390 KIAS in a stabilized line abreast position where both aircraft deployed the speed brakes and slowed to 300 KIAS. With the MiG-17F at 300 KIAS, the F-100 still had 340 KIAS and was more than 1,000 feet in front of the Fresco C.

The exploitation team employed offensive and defensive maneuvers against the Fresco C on all F-100 flights. Offensively, the MiG could force the F-100 into overshoot, regardless of his plan of attack. Defensively, the F-100 effectively used afterburner accelerations to depart a vulnerable position. Afterburner acceleration by the F-100, however, could not immediately produce safe separation while subjected to gunfire.

Preventing tracking while achieving separation required some last-ditch escape maneuver. Unloaded reversals at high-calibrated airspeeds during flight 24 proved very effective in producing that result. In summary, the F-100 should pursue an attack on the Fresco C until tracking was no longer possible, maintaining a high attack airspeed.

Defensively, the F-100 employed afterburner accelerations to defeat Fresco C gun attacks. If detection occurred after the Fresco C had opened fire, the F-100 should maneuver to destroy a tracking solution while accelerating with an unloaded acceleration complemented by rapid reversals at random intervals. If altitude permitted, an escape maneuver should progress toward a steep dive to force the

Fresco C into a high-speed flight if he chooses to follow the maneuver with reattack decisions based on tactical considerations. A reattack by the F-100 risky if a competent Fresco C pilot elected to pursue.

The team did not evaluate flight tactics to employ against the Fresco C when the F-100 enjoyed numerical superiority. However, it became apparent that the F-100 could effectively employ split plane maneuvering and mutual support to defeat the Fresco C in a two-on-one situation.

The Following Conclusions Pertain to Findings Derived from Competition Between a Clean F-100 And A Clean Fresco C.

With both aircraft operating at military power, the F-100 possessed an acceleration capability that slightly exceeded that of the Fresco. Conducting an acceleration comparison under unloaded conditions showed the two aircraft having comparable acceleration potential.

The acceleration capability of the F-100 exceeded that of the Fresco C with both aircraft operating at maximum power.

The zoom capability of the F-100 was comparable to that of the Fresco C, whether conducted at maximum power or military power.

At 350 KIAS or below, the Fresco C had a turning capability superior to that of the F-100, and below 25O KIAS, the Fresco C was far superior.

At airspeeds approaching 450 KIAS, the turning capability of the F-100 was comparable to that of the Fresco C, and perhaps slightly better; however, the Fresco C lost far less speed in the turn. The Fresco C regained turning superiority by using a speed brake deceleration in the turn.

At comparable power settings, the Fresco C decelerated more rapidly with speed brakes than did the F-100.

The roll rate of the F-100 was superior to that of the Fresco C.

From a position of advantage, the F-100 employed a hit-and-run maneuver to defeat the Fresco C.

When defensive, the F-100 negated an attack by the Fresco C through the execution of acceleration to supersonic speed. However, it did so with less the facility than did the F-4 or F-100. Failure of acceleration alone negating an attack by the Fresco C, required a last-ditch maneuver when possible, progressing to a steep dive.

Exploitation results concluded that, following an initial Fresco C confrontation and subsequent separation, crews avoid attempting to position for reattack in a one-on-one situation, unless they enjoy a tactical advantage.

MiG-17F FRESCO C vs. F-5

F-5A versus MiG-17F Fresco C

Missions 22 and 23 evaluated the F-5A in a clean configuration in the presence of a Fresco C for acceleration comparisons. At military power, originating at 250 KIAS, the F-5A accelerated to 400 KIAS and was some 3,000 feet in front of the MiG-17. The MiG had 360 KIAS When the same comparison occurred in unloaded flight; the F-5A did not obtain a significant advantage until approximately 350 KIAS. Maximum power illustrated a distinct advantage for the F-5, which accelerated from 300 to 500 KIAS while the MiG-17F accelerated from 300 to 410 KIAS.

Zoom comparisons occurred on both flights, one each at maximum and military power. At military power, the F-5 zoomed 1,800 feet above the Fresco C, which terminated 30 knots faster. At maximum power, the F-5 zoomed 2,500 feet above the Fresco C and was 80 knots faster at termination.

Turn comparisons occurred at 450 and 350 KIAS, maximum and military power. At 450 KIAS, the F-5 completed a 180-degree turn sooner than the MiG-17, and at 350 KIAS, the two aircraft demonstrated comparable turn rates and radii. In all cases, the F-5 lost considerably more airspeed in the turns than did the Fresco C. Subsequent offensive and defensive maneuvers illustrated to the pilots of both aircraft that 300 KIAS approximated the airspeed below which the Fresco C enjoyed a definite turning advantage.

Roll rate comparisons conducted on flight 23 disclosed a wide margin of superiority for the F-5A at

350 KIAS and 450 KIAS. At 350 KIAS, the F-5 rolled 360 degrees, and the MiG-17 rolled 180 degrees. At 450 KIAS, the F-5 completed a 360-degree roll simultaneously with the completion of 90 degrees of the roll by the Fresco C.

The MiG flew speed brake comparisons on both F-5 missions, one at constant power, and one at idle power. The constant power deceleration initiated at 400 KIAS and terminated when the MiG-17 indicated 300 knots. Now, the F-5 had 320 KIAS and was approximately 2,000 feet in front of the Fresco C. At idle power, the advantage of the Fresco C was less apparent, and over a 130-knot deceleration range, the F-5 was only 500–800 feet in front at termination.

During offensive and defensive maneuvers on three occasions, the F-5 offensively tracked a level breaking Fresco C long enough to obtain a good gun burst when the Fresco C airspeed was above 300 KIAS. If the MiG-17 executed an oblique vertical hard turn, however, it rapidly bled airspeed and generated an F-5 overshoot before it adversary could accomplish effective gun-tracking.

Defensively, the F-5 could not force an overshoot with a hard/break turn. As with all aircraft testing on this evaluation, the F-5 exploited its superior maximum power acceleration potential to prevent a detected threat from arriving at a lethal gun position.

In summary, in a one-on-one situation, the F-5 pressed an attack from a position of an advantage until tracking the turning Fresco C became impossible. At that point, the F-5 executed a level or unloaded acceleration. Separation ultimately occurred, and the F-5 retained a high attack speed to facilitate separation. The point at which to execute such a maneuver constituting a difficult decision for the F-5 pilot unaccustomed to engaging aircraft with which they cannot turn below 300 KIAS.

Defensively, the F-5 accelerated to defeat a Fresco C gun attack. Detecting the attack late required a last-ditch track spoiling maneuver, such as an accelerated, unloaded roll. Such an escape maneuver carried into a steep dive to force the Fresco C into an undesirable speed realm if conditions so permitted. It was worthy to note the F-5 enjoyed an advantage uncommon to the F-100, F-4, and F-105, insofar as it, like the Fresco, was difficult to see.

During separation, it became apparent when the enemy lost visual contact, at which time re-attack represented a judicious course. The F-5 best positioned for a re-attack with a supersonic, low g zoom to a position of advantage.

The exploitation team did not evaluate flight tactics employed by the F-5 against a Fresco C threat. However, it seemed logical that the F-5, when numerically superior, defeated the Fresco C by using split plane maneuvering and mutual support. Such tactics judiciously applied prevented the Fresco C from achieving a lethal position while the free element brings offensive pressure to bear, thus permitting the F-5 to control the engagement.

The following conclusions pertain to a clean F-5A and clean Fresco C.

With both aircraft operating at military power, the F-5A possessed an acceleration capability superior to that of the Fresco C.

With both aircraft operating at maximum power, the F-5A possessed an acceleration capability, superior to that of the Fresco C.

The F-5A zoom capability was superior to that of the Fresco C.

Below 300 KIAS, the turning capability of the F-5A was inferior to that of the Fresco. At 350 KIAS, the turning capability of the F-5A was comparable to that of the Fresco C, and at 450 KIAS, the F-5A possessed a better turning capability had Fresco. In all cases, however, the F-5A lost considerably more airspeed in the turn. In the latter two cases, the Fresco C regained turning superiority or, at worst, reduced the margin of inferiority by employing a speed brake deceleration in the turn.

The F-5A could not decelerate as readily with speed brakes as could the Fresco C.

The F-5A rolled at a rate far exceeding that of the Fresco C and at a vastly superior rate at 450 KIAS and above.

From a position of advantage, the F-5A best defeated the Fresco C through the employment of hit and ran maneuvers.

When defensive, the F-5A defeated an attack by the Fresco C through acceleration to supersonic speed.

When the F-5 was defensive and the Fresco C too close to deny it a lethal position by accelerating alone, the pilot accomplished a last-ditch maneuver to defeat that position. Upon achieving this goal, the F-5A immediately separated, using a steep dive.

The Fresco C flew at its maximum performance capability throughout this evaluation. The following on was against a single Fresco in contrast to anticipated multiple Fresco Cs in combat. Their exploitation results concluded that: flight tactics employed by USAF tactical fighters against Fresco C type aircraft emphasized mutual support between elements and split plane maneuvering. USAF tactical fighters employ the superior acceleration and vertical maneuvering potential at their command to control engagements with the Fresco.

Future tactical fighters should consider small or negligible exhaust trails a primary engine design consideration.

Review and update all wing tactical fighter doctrines dealing with specifics concerning Fresco C engagement criteria to include the findings of this test.

A portion of combat crew training, both initial and recurrency, be devoted to a comprehensive instruction concerning aircraft with various performance characteristics.

Continue expending effort in the future acquisition and tactical exploitation of foreign material.

USAF tactical fighters use steep dive angle (30 to 90 degrees) escape maneuvers when disengaging the Fresco C if circumstances permit.

USAF tactical fighters keep ingress speed sufficiently high to permit rapid acceleration to negate a Fresco C gun attack from the rear hemisphere.

Fighter aircrews remain acutely aware of the visual acquisition problem presented by Fresco C size targets before, during, and after an engagement.

F-105 Versus MiG-17-F Fresco C

Fresco C with Normal
Russian Markings

The Air Force and Navy conducted zoom comparisons on flights 7, 14, 28, and 40. With both aircraft using maximum power in the profile described in Test Environment and Procedures, the F-105 averaged a 4,500-foot and 30-knot advantage at termination. The military power zoom was less favorable to the F-105, however, it still had some 1,000 feet in its favor. On that maneuver, the F-105 had 20 knots less at termination. During a single zoom comparison with the MiG-17 at maximum power, and the F-105 at military power, the MiG-17F terminated 500 feet above and 200+ knots faster than the F-105.

Turn comparisons performed on flights 5, 6, 12, and 50 demonstrated that at 350 KIAS, the Fresco C could turn far better than the F-105. At Mach .9 or 450 KIAS, the results varied, depending on the power settings used. At military power, the Fresco C enjoyed a slight advantage, and at maximum power; the F-

105 turned slightly better in all cases. However, the MiG-17 retained more energy and terminated at a higher airspeed.

At 350 KIAS, during roll rate comparisons performed on flight 7, the two aircraft rolled at approximately the same rate. At 450 KIAS, aerodynamic forces opposed the deflection of the Fresco C flight controls. Consequently, the F-105 rolled out of a 360-degree aileron roll before the MiG-17 had reached an inverted position.

Speed brake deceleration comparisons performed on flights 5 and 50 indicated at average an approximate 100-knot deceleration range; the Fresco C decelerated 15 to 20 knots more than the F-105.

The exploitation team collected data from the RHAW, radar homing and warning receiver, pertinent to the range only radar of the MiG-17F, and as previously mentioned in the F-4 portion of the discussion. The electronic characteristics of the radar of the Fresco C illuminated the AAA/AI light of the APR-25/26. The differentiation between that indication and a gun-laying radar indication was difficult in a multi-signal environment. AAA/AI indications underwent careful analyzing before attributing it to an anti-aircraft artillery radar.

An evaluation of F-105 offensive and defensive maneuvers employed against a Fresco C occurred on flights 5–19 12, 14, 28, 29, 40, 41, 49, and 50. Offensively, the F-105 employed yo-yos, lag pursuit attacks, and maneuvers in the vertical plane in attempting to defeat the Fresco. Hard break turns by the Fresco C, however, invariably forced the F-105 to overshoot and sought separation.

Defensively, the F-105 could not generate an overshoot through hard/break turns and resorted to other means to protect itself.

As with the F-4, the exploitation team found the superior maximum power acceleration potential of the F-105 provided it with the best defensive maneuver against the Fresco. If, however, detection of the MiG-17 positioned in a lethal position required some last-ditch maneuver for safety during separation.

An unloaded acceleration with subsequent high-calibrated airspeed reversals was the most effective means to that end. Flights 29 and 41 made it apparent the pilot should keep ingress airspeeds sufficiently high to produce rapid acceleration in the presence of a MiG threat.

In summary, in a one-on-one situation, the F-105 should pursue an attack until tracking becomes impossible, and seek separation immediately, keeping airspeed high throughout the attack. The F-105 cannot afford to engage the Fresco C at a high angle of attack, slow speed, turning fight.

Defensively, the F-105 employed an afterburner acceleration to nullify an attack by the Fresco C in an unloaded flight if the threat was from close in. Permitting an escape maneuver to progress into a steep dive angle benefited the F-105 since it enhanced acceleration as well as forced the Fresco C into high-speed flight. High calibrated airspeed reversals at random intervals effectively produced safety while generating separation. After separation, the F-105 should use a very low g turn back to observe the subsequent actions of the Fresco. If the threat follows, reattack will not normally be judicious.

F-105 flight tactics against the Fresco C evaluated on flights 12, and 14 used split plane maneuvering and mutual support, bringing constant pressure to bear on the Fresco C. Both elements proved more effective when maintaining high-calibrated airspeeds.

If the F-105 crew could avoid the temptation to track a breaking Fresco C and separate when an overshoot was inevitable, the flight tactics described could prevent the Fresco C from ever acquiring a lethal position. Encountering a single Fresco C required maintaining a constant vigil for another aircraft, and mutual support included constant clearing of the rear hemisphere between elements.

Conclusions Pertinent to an F-105 Aircraft Configured With AIM-9 Pylons on The Wing Stations and Simulated Electronic Countermeasures Pods, And the Fresco C Clean

- With both aircraft operating at military power, the F-105 had an acceleration capability superior to that of the Fresco C above 250 KIAS.
- With both aircraft operating at maximum power, the acceleration capability of the F-105 was superior to that of the Fresco C.
- With both aircraft operating at military power, the zoom capability of the F-105 was slightly superior to that of the Fresco C.
- With both aircraft operating at maximum power, the F-105 possessed a zoom potential

superior to that of the Fresco C.

- At 350 KIAS or below, the Fresco C possessed a turning capability superior to that of the F-105 and the margin of superiority increases with a corresponding/decrease in airspeed.

As airspeed approached 450 KIAS, the turning capability of the F-105 approached that of the Fresco C. The F-105 lost considerably more airspeed for the same heading change. Further, the Fresco C could easily regain turning superiority by decelerating in the turn with speed brakes,

When both aircraft employed comparable power settings, the Fresco C decelerated faster with speed brakes than could the F-105.

The F-105 roll rate with full-scale stick deflection was superior to that of the Fresco C at high-calibrated airspeeds and comparable at 350 KIAS or below.

From an initially advantageous position, the F-105 should use a hit and run maneuver to defeat the Fresco C.

Defensively, the F-105 could defeat an attack by the Fresco C by accelerating to supersonic speed. At too close range to permit immediate safety through acceleration, the F-105 employed a last-ditch maneuver to defeat the lethal position of the Fresco. Seek separation, using a steep dive angle, if possible. The F-105D and F-105F opposed the MiG-17F MiG-17 on flights 5–7, 12–14, 28, 29, 40, 41, and 50. Configurations for the F-105 varied as required to fulfill flight objectives with basic comparison data compiled with the F-105F configured either clean or with AIM-9 rails on the wing stations.

They conducted acceleration comparisons in level and unloaded flight and at maximum and military power during flights 5, 6, 28, 41, and 50. At military power, the F-105 possessed an acceleration capability exceeding that of the Fresco C. However; this was not decidedly so until near 275 to 300 KIAS.

At maximum power, the F-105 enjoyed a decisive acceleration superiority over the Fresco C.

The Fresco C had a maximum power acceleration capability that slightly exceeded the military power acceleration capability of the F-105.

Fresco C and F-105

Chapter 9 - MiG-17 USN Tactical Evaluation

It is significant to note the Navy lost every simulated air combat exercise on their first sortie against the HAVE DOUGHNUT MiG-21 Fishbed, HAVE DRILL, and HAVE FERRY MiG-17F Frescos.

Navy A-4, A-6, A-7 air attack aircraft failed to gain the offensive on the Fresco C in any flight regime throughout the tactical phase of the named exploitation projects. At CRT, the Fresco C out accelerated, out climbed, and out zoomed all Navy attack aircraft Below 475 KIAS, the Fresco C enjoyed a marked advantage in sustained turn performance. Maximum steady state roll rate was the only area of advantage for the attack planes. The Fresco C excelled in the slow speed regime. The A-7 made two reversals in a scissors maneuver before the Fresco C reaching a gun-tracking position. The A-4 and A-6 completed only one reversal before losing the slow speed fight. None of the attack aircraft gained an offensive position on the Fresco C.

The Navy attack aircraft A-4, A-6, A-7 air attack aircraft successfully disengaged from the Fresco. The superior roll rate and better high j maneuverability of all Navy attack aircraft permitted them to disengage and escaped from the Fresco C.

With the Fresco C in a gun-tracking position, unloaded maximum roll rate reversals, while accelerating to maximum q, negated the Fresco's tracking solution immediately and placed the attack aircraft outside the maximum gun range.

Of advantage to the attack, aircraft pilots were their familiarity with high-speed low-altitude maneuvering. Because of the flight control problems, the Fresco C pilots felt extremely uncomfortable attempting to follow attack aircraft down to treetop level at speeds exceeding 475 KIAS.

The Fresco C pilots usually refused to follow the attack aircraft into this regime. The A-6, with a higher q limit, escaped with more facility than either the A-4 or A. Once the Fresco C achieved a gun-tracking position, a breaking turn by the attack aircraft failed to cause an overshoot.

Losing 100% of their first encounters against the MiG in Nevada turned the Navy's focus to human factors and personnel evaluations. It is important to note that overconfidence was an important derogating factor in the performance of US Navy pilots while fighting the Fresco C for the first time. Before fighting the Fresco, there was a tendency to underestimate its performance. Although last produced in 1961, the Fresco was still a formidable threat to all Navy tactical aircraft

Aircrews remained constantly vigilant and not lulled into complacency by the age of the Fresco. Pilot performance improved dramatically after the first engagement with the Fresco. All project pilots agreed that before their first engagement; they had seriously underestimated the capabilities of the Fresco.

Though subsonic, the Fresco was a dangerous adversary. Substantial information concerning the actual capabilities of the Fresco C was available before Project HAVE DRILL. In all significant cases, the results of the project validated this information. Underestimating the Fresco was due to its vintage 1950 and simplicity as compared to its opponent, i.e., F-4, F-8, etc., and not the published performance characteristics.

Due to system simplicity and the absence of dangerous flight characteristics, the pilot employed the Fresco to its fullest extent, finding the aircraft easy to fly. The pilot could not exceed the structural limit load factor of 11 g. The lack of abnormal stall or spin characteristics allowed the aggressive pilot to approach the edges of the Fresco's flight envelope with confidence.

The weapons system simplicity eliminated much of the pilot confusion presented when using systems that are more complex. After minimum training, the pilot realized that he could not overstress the aircraft, could quickly recover from the uncontrolled flight, and effectively employ the weapons system. This overall simplicity and ruggedness became of great significance in the ACM environment.

Like the Air Force, Navy attempted with its flights to duplicate the ACM environment encountered in SEA. However, unlike true combat, the Navy choreographed its exploitation program with the other participants. The Navy briefed the participating pilots before each flight, maintained two-way UHF radio communication among participating pilots, and terminated engagements when encountering unusual flight characteristics.

The participants strived to assume a covering Fighter/Engaged Fighter posture in every engagement. If attacked from aft of abeam while in combat spread, they unloaded for airspeed and separation (500 KIAS minimum). They increased the vertical and lateral separation between their aircraft by the outside man going up and outboard, applying g only to establish a climbing attitude. If the MiG-17s split, the Navy pilots continued to accelerate and extended the fight beyond their q limit.

When approaching 600 KIAS, the inside pilot started a slight turn away from his wingman who then maneuvered for an attack on the inside MiG-17. The separation between the outside friendly and the outside MiG combined with the high airspeed prevented his reentering the fight before his wingman came under attack.

Fresco C w/USN Attack Airplanes

If the Frescos choose to stay together on one friendly, the Navy pilots dragged them out to their q limit while increasing the vertical and lateral separation between fighters. At this point, the engaged fighter (the one the MiGs chose to chase) performed a flight turn away from his wingman who turned in for his attack as the MiG started to follow him around. The covering fighter avoided bleeding off speed to below 475 KIAS and maneuvered into the MiG-17's blind area, i.e., below the plane of the MiG-17's wings.

If the Frescos decided to slow down and turn, the covering fighter unloaded and separated, attempting to drag the MiGs after him. At this point, he called his wingman in to assume the covering fighter role.

If attacked by a MiG-17 one-on-one, they did not try to force an overshoot. Instead, they sacrificed angle off to gain nose to tail separation, while accelerating beyond the Fresco's q limit, jinking while accelerating to destroy his gun-tracking solution. Jinking meant changing the flight path of the aircraft in all planes at random intervals.

During separation, they maintained visual contact and avoided arcing excessively to prevent the Fresco C cutting across their circle and getting a shot. As the range approached visual detection limits (2–3 nautical mile), they performed a maximum rate, minimum radius reversal to meet the Fresco head-on while striving for maximum efficiency of their plane. Knowing the Fresco's performance contingent on the use of afterburner, the Navy stressed winning the fight by having fuel left when the MiG pilot bingoed.

The Navy's strategy exploited the Fresco's slow roll rate and blind cone below the plane of his wings, requiring the Fresco pilot to counter with a roll where the MiG-17's poor roll rate worked against it. Knowing the MiG-17's performance declared severely above 475 KIAS, the Navy fighter planes attempted to fight the MiG-17 at energy levels exceeding the capabilities of the MiG-17.

The A/B on the Fresco engine gave it a performance level impossible to duplicate or realistically simulate by US aircraft with similar turn capability. Thus, even with 2 1/2–3 nautical mile separation, it required lots of technique to turn for reengagement.

If an aggressive Fresco pilot saw an attack by Navy fighters in the aft hemisphere, he could force an overshoot at all airspeeds below 475 KIAS. The Navy fighters learned not to slow down and turn with the Fresco, executing a high yo-yo when practicable.

A roll away during the yo-yo was an effective lag pursuit maneuver and put the attacking fighter in an area difficult to see (i.e., high, and aft) and one where the Fresco pilot did not expect the fighter to be.

This method of countering the overshoot normally increased the longitudinal separation between the Fresco and fighter and, thus, aided in weapons system employment. The reversal at the top was normally a rolling split-S (bank angle about 135 degrees). Most importantly, this method helped preclude trapping the fighter into a very close-in turning duel.

Classic Loose Deuce maneuvering was a great aid in maintaining mutual support with the greater separations required to combat a section of Frescos effectively. Aircrews pressing for the "quick kill" made mutual support challenging and ran the risk of allowing the fight to degenerate into a turning duel.

The significant performance advantage enjoyed by the F-8 and F-4 over the Fresco eliminated the need to use full A/B. Afterburner was not required as frequently or for such great intervals of time, as when fighting a Fishbed or other similarly performing aircraft. Fuel was one of the Fresco's prime constraints and severely limited its offensive capability.

The tactical engagements over Groom Lake established that A-4, A-6, and A-7 aircraft should not attempt to engage the MiG-17 in ACM. If engaged, they should escape expeditiously by using a superior roll rate, unloaded accelerations, and high-speed, low-level run outs.

When a sighting a Fresco attacking you, go to full power and immediately jettison all external stores while creating as much angular separation as possible while staying unloaded and diving for airspeed.

If the Fresco closed to a gun firing position, perform a maximum rate roll away from him, staying unloaded and continue your acceleration. Hold this position for 1–2 seconds and then roll at maximum rate back toward the Fresco.

If you kept g off the aircraft and continued your unloaded diving acceleration, the range opened significantly. Take the Fresco down to as low a level above the ground as possible, maintaining your highest airspeed. The Fresco pilot was unable to attack you, low-level, at speeds above 500 KIAS.

If necessary to engage the Fresco, use the fighter offensive separation tactics, described earlier in this section to exhaust the Fresco's fuel supply. Take extreme care under these circumstances to avoid other enemy aircraft that could choose to enter the fight.

It was the opinion of the MiG-17 pilots that the vulnerability to combat damage of the Fresco was quite low. They found the Fresco a rugged and durable aircraft Since the VK-1F was a centrifugal flow engine, its vulnerability to gunfire or fragment damage was much lower than that of the axial flow engine. The MiG-17 only hydraulically boosted the ailerons and provided manual backups with mechanical pushrods provided for all controls, which also lowered the kill probability from gunfire or fragmentation. It contained all its fuel cells in the fuselage, making the wings relatively invulnerable. Armor plating protected the pilot.

The Navy concluded the Fresco C capable of defeating any USN tactical aircraft in a turning fight at speeds of 475 KIAS and below. In the subsonic region, the Fresco C, properly employed and aggressively maneuvered, was a highly effective air superiority weapon system.

The Fresco C was straightforward and reliable enough to operate from remote sites with a minimum of support equipment. They found the subsonic maneuvering capability and overall performance of the MiG-17 grossly underestimated. One could fly the MiG-17 with confidence to the limits of its operating envelope.

The gun system of Fresco C was highly reliable and effective The Fresco C had a lower vulnerability to combat damage than most current US Aircraft. The Fresco C had good visibility above its horizontal plane poor visibility below its horizontal plane. Camouflage paint and the small size of the aircraft made a visual acquisition and retention of the Fresco C difficult in the air-to-air environment. Its fuel limitations prevented the Fresco C engaging in prolonged ACM beyond 75 m of its home base.

The q limit of the MiG-17 seriously inhibited its total effectiveness in ACM. The Fresco C was extremely fuel limited. All USN tactical aircraft had higher q limits than the MiG-17. All USN tactical aircraft had higher roll rates than the MiG-17.

The head-on VID as currently flown in the 1–4 was not effective against the Fresco because it was dependent on full system radar track by both attacking aircraft and sacrifices tactically advantageous positioning. USN attack aircraft had no offensive capability against a properly flown MiG-17.

The oblique loop maneuver used against the MiG-21 was ineffective and dangerous against the MiG-17. Due to overall performance superiority, two properly flown F-4s could remain 100 percent offensive against two MiG-17's in air combat. The Padlock technique was mandatory during ACM with the MiG-17.

MiG-17F Fresco Exploitation Conclusions

Navy Recommendations

Commander Operational Test and Evaluation Force recommended that: Navy fighter aircraft use the following tactics to defeat the Fresco C.
- Maintain a high-energy level-500–600 KIAS.
- Avoid high g maneuvers below 500 KIAS.
- Use thrust advantage to prevent the Fresco from attaining a lethal gun-tracking position.
- Force the Fresco to fight at airspeeds above 475 KIAS.
- Engage only as a section with strict mutual support, using "loose Deuce" maneuvers and striving to maintain an offensive fighter posture.
- Use A/B judiciously and efficiently to take maximum advantage of the Fresco's limited fuel supply.
- Exploit the Fresco's weaknesses, i.e., the blind area below the horizontal plane, poor roll rate and marginal control ability in the high q regime.
- Use when necessary, maximum rate, minimum radius turns at ranges to the Fresco of 2–3 nautical miles, and base the decision to reverse on respective energy levels and the tactical

situation.

- A-4, A-6, and A-7 aircraft do not engage in ACM with the Fresco C.
- If engaging the Fresco C with A-4, A-6, and A-7 aircraft, the pilot should disengage by jettisoning all stores not possessing air-to-air capabilities. He or she should perform an unloaded acceleration while diving to low-level, by using maximum roll rate and running out at maximum airspeed and minimum altitude.
- If a reversal or than was necessary, ensure that the range to the Fresco was greater than 2–3 nautical miles, and then perform a minimum radius and maximum rate turn to pass the Fresco head-on.
- Maintain strict lookout always in hostile territory.
- In a threat area, weave and vary headings along the basic course•
- Practice realistic AC25 as much as possible under controlled conditions against small aircraft with low wing loading.
- Whenever possible, conduct radar intercept and training over land.
- Pilots practice coaching RIO's on to targets always, even when not engaged in ACM.
- Intensify squadron level training of RIOs in air combat tactics.
- Intensify ACM training of attack aircrews at all levels.
- Pursue continued exploitation of foreign aircraft

Tactical flight Summaries

Summarizing the HAVE DRILL and HAVE FERRY exploitation efforts, the exploitation found the MiG-17F easy to fly, and the undesirable features, such as spin and accelerated stall, readily handled when encountered. With proper user knowledge of its maneuvering capabilities and limitations, the exploitation team found the MiG-17F a very effective interceptor/air superiority daylight fighter throughout most of the subsonic flight envelope.

The MiG-17F demonstrated outstanding lift-limited maneuvering capabilities. Available g and g-level for onset buffet ran high and turn radius low. Thrust limited turning performance was also good. Longitudinal stick forces increased significantly beyond .85 Mach number and remained excessive above .90 Mach. Lateral-directional damping was weak and especially poor near .92 Mach Number Lateral-directional tracking for gunfire suffered from turbulent weather or small, inadvertent control disturbances,

Roll rate capability was low, only 100–130 degrees per second. Sufficient aileron deflection was not available past 572 KIAS to prevent the aircraft rolling off.

The exploitation team found the MiG-17F very reliable and its operational availability exceptional. The exploitation team easily accomplished four to five flights day after day with briefings, debriefings, and turnaround times of other participating aircraft the only factors limiting additional flights. Considering these conditions, the reliability record of the MiG-17F, during the evaluation, was not only exceptional, however, was a sobering fact.

The MiG-17F was a very easy aircraft to maintain. Soviet design approached a rugged airframe, engine and system simplicity and overall reliability during the manufacturing period of this plane. Attractive features included dependable in-flight gun clearing and charging, and a smooth, well-balanced flight control system. Particularly impressive was the lack of engine exhaust smoke at lower altitudes.

The MiG-17F's simplicity and rugged design lent it the capabilities to operate from unprepared landing strips and for full maintaining by underdeveloped nations, thus allowing it a continued threat during limited war situations.

The MiG-17F design concentrated the vulnerable critical flight components in a small volume thus presenting a small, vulnerable target. The survivability design philosophy of this aircraft protected these vulnerable components by masking and armor from the front and rear. Vulnerability to continuous rod warheads, conventional armaments used in the AIM-7E, AIM-90, and AIM-54A, was slightly lower than previously assumed by the ordnance community. The factors pointing to this assessment were the liberal use of high-strength steel components throughout the aircraft and the push-pull rod system used for control actuation.

The VK-1F engine of the MiG-17F performed close to factory specification during the HAVE DRILL program. Exhaust gas temperature, thrust, and fuel flow stayed within the manufacturers prescribed operating limits. The maximum physical RPM at which the exploitation team ran the engine was at an average 1.12 percent below a specified value. Engine pressure ratio at military power was in exact agreement with the Soviet published data.

The reader should note the author basing the performance data presented in this book upon the individual aircraft studied. The author attributes differences between test data and published estimates to nonstandard day conditions, uncorrected engine performance, and incorrect comparisons between flight rules or profiles. In most cases, comparisons appeared representative of typical aircraft flight-testing results using all sources of information for correctly and analyzed the weapon system capabilities completely.

Both programs wound down by June 1969, with the findings shared with the Navy's new TOPGUN School established after these exploitation programs revealed the need to reintroduce dogfighting skills to Navy pilots.

The Navy shared the results of HAVE DOUGHNUT, HAVE DRILL, and HAVE FERRY with the instructors at the USAF's Fighter Weapons School at Nellis AFB, Nevada. This later induced the Air Force to establish the RED FLAG aerial combat training exercises mimicking real air combat situations.

Continue USAF efforts to acquire threat type fighter aircraft for exploitation.

This project provided urgently required information for the Air Defense Command with a unit presently deployed to Korea where the MiG-17 posed most of the aerial threats.

The results of this project left the unit and future deploying units better prepared to complete their flights. Information gained from "HAVE DRILL, HAVE FERRY," and future foreign aircraft exploitation ensured the development of adequate tactics to counter those aircraft and design of future US fighters with superior performance, systems, and armament. Aerospace Defense Command remained prepared to participate in the exploitation of foreign aircraft to ensure the combat effectiveness of US military forces.

Summary of Delayed Discrepancies and Required Maintenance Generated Because Of The Project.
- VHF radio removed
- IFF not installed
- Glare shield bracket removed
- VHF power supply removed
- All ammunition cans removed
- VHF control head removed
- Radio compassed indicator removed
- Left wing IFF crystal removed
- Ballast installed in nose bay (84 lbs).
- Right wing IFF crystals (2 ea,) removed
- Tactical bombing system inoperative, 10 AMP fuse removed
- SIRSITA bell in cockpit removed
- Afterburner flame holder distorted and cracked
- APT section removed to facilitate other maintenance
- Afterburner assembly removed
- Two holes melted through the heat shield in APT section
- Bottom afterburner eyelid actuator APT mount chaffing AFT section
- Hydraulic line to bottom afterburner eyelid actuator chaffing AFT section
- Bottom rib in the APT section below the bottom afterburner actuator cracked and distorted
- Brakes weak

Maintainability (HAVE DRILL)

The following figures presented the man-hours expended for maintenance on the HAVE DRILL MiG-17 along with some sortie information.

- 498.9 Man-hours for discrepancies
- 89.9 Man-hours for periodic inspections
- 86.0 Man-hours for post flights and preflights
- 674.8 Total man-hours for maintenance
- 131.3 Total flying hours. 172 Total sorties
- 5.1 Man-hour per flying hour 3.9 man-hour per sortie
- 101 Maintenance discrepancies of these, 30 with the radar
- 498.9 Man-hours for discrepancies of these, 259–7 M.H. with the radar
- 55 Flying days. 3.1 Sorties per flying day

The aircraft turned around several times in 30 minutes. Including 10 minutes towing time at the end of the runway before takeoff

-

MiG-17 with F-4 and F-8

The MiG-17F HAVE DRILL & HAVE FERRY WITH NAVY F-8

The exploitation team discovered the last eight discrepancies when the project terminated.

The Fresco C encountered aerodynamic forces that oppose deflection of the flight controls above Mach .85 or 450 KIA, resulting in a very slow available roll rate and pitch change in that speed realm.

The Fresco C flight controls lacked dampening in any axis. Above 375 KIAS, the aircraft possessed a Dutch Roll tendency. At any speed in turbulent air, the aircraft was difficult to control in yaw.

The Following Deficiencies Limited the Fresco C.
- Low firing rates of the 2 mm, and 37 mm cannons (900 and 400 rounds per minute, respectively).
- Low muzzle velocity of the 23 mm and 37 mm ammunition (2,250 fps for both calibers).
- Excessive tracking time for proper lead computation (two to three seconds).
- Total lead required due to low muzzle velocities and projectile weights.

The exploitation team concluded from these flights that flight tactics employed against the Fresco C by all tactical fighter aircraft that participated in this evaluation should emphasize mutual support between elements and split plane maneuvering.

The prolonged high angle of attack maximum performance maneuvering and low airspeed (250 KIAS or slower) defined a performance envelope wherein the Fresco C was superior to all tactical fighters that it confronted with this test.

The F-4, F-105, F-100, and F-5 aircraft all possessed a vertical maneuvering superiority over the Fresco. The F-4 and F-5 aircraft provided a decisive advantage with low g pitch movements to the vertical plane.

When the Fresco C engaged any of the tactical aircraft at calibrated airspeeds exceeding 450 KIAS, USAF aircraft defeated its lethal position by using rapidly unloaded reversals and check turns of 30 to 60 degrees while maintaining high-calibrated airspeeds.

USAF tactical fighters could accelerate in low nose attitudes of 30 to 90 degrees during an escape maneuver, forcing the Fresco C pilot into a realm of flight wherein his capability to pull out becomes his dominant consideration.

The F-4 was by far the easiest of the participating aircraft to acquire visually, followed by the F-105 and F-100, in that order. Both the Fresco C and F-5 proved very difficult to acquire with the unaided eye. Most first encounters with the Fresco C demonstrated that highly qualified pilots made serious misjudgments in range estimation.

Incorporate the findings of this test into the tactical doctrines of Southeast Asia fighter wings for more accurate instructions for Fresco C engagements.

The APR-25/26 of the F-4 and F-105 furnished an indication of the presence of the Fresco C-Scan Fix, range-only radar that accompanied the strobe by the illumination of the AAA/A1 light.

The MiG Fresco C aircraft was relatively easy to fly and required minimum time to obtain an average proficiency level.

At medium and low altitudes, the aircraft appeared to operate best at speeds approximating 300 to 350 KIAS.

Its acceleration potential of low speeds (180 KIAS or less) to 250 KIAS provided a significant advantage to the Fresco C in a low-speed maneuvering engagement.

At airspeeds of 350 KIAS or below, the Fresco C possessed an outstanding turning performance.

Above Mach .85 or 450 KIAS, whichever was lower, the Fresco C encountered aerodynamic forces that opposed the deflection of the flight controls; hence, in that speed realm, roll, and pitch rates deteriorated markedly.

The aircraft was extremely small and difficult to acquire visually, a characteristic enhanced by a negligible exhaust trail.

The maximum speed of the aircraft at low and medium altitude in level flight was approximately Mach .96. High stick forces prevented maximum performance maneuvering in that speed range.

The bubble canopy, with a mounted periscope assembly, afforded the Fresco C excellent visual acquisition in the rear hemisphere. Poor visibility looking low, both forward and laterally, and the periscope assembly interfered somewhat with overhead visibility.

When configured clean with full internal fuel, the Fresco C had approximately 373 gallons °F fuel available to the engine. Maximum performance maneuvering at medium or low altitude consumed the available fuel available in 20 to 25 minutes.

The Fresco C flight controls lacked dampening in any axis. In turbulence, control of the aircraft was difficult in the yaw axis. At 375 KIAS or above, it demonstrated Dutch Roll tendencies or instability in all axes.

When the Fresco C employed against aerial or ground targets, the weapons system was effective under ideal conditions. However, low cannon firing rates, low projectile muzzle velocities, excessive lead computation time, the excessive requirement for the lead, and excessive yawing motion in turbulent air limited the overall effectiveness of the weapons system.

TAC Summary of The Test Program of The MiG-17F Fresco C

Before starting the tactical phase of exploitation, the two primary TAC pilots flying most the total TAC flights (52 of 57) checked out in the Fresco C. They performed this in two flights and one flight, respectively before the initiation of the TAC test program. A third TAC pilot received a two-mission checkout program on TAC flights 38 and 39. All three TAC pilots expressed the following opinions regarding the operational characteristics of the MiG-17:

- They encountered no difficulty in starting the aircraft
- The aircraft was somewhat cumbersome to taxi. This fact arose, in part, from the peculiar mechanics involved in braking, and utilization of differential braking for directional control. Applying hand pressure to a level on the control stick column controlled air brake pressure, and deflecting the rudder pedals accomplished differential braking.
- Once airborne, the aircraft exhibited characteristics that rendered the pilot comfortable at the controls. All three pilots felt capable of operating the aircraft at maximum performance after concluding their initial checkouts.

Throughout most the flights, 300 to 350 KIAS appeared as the optimum airspeed range for maneuverability and weapons employment. (Reference flights 1–9, 26, 27, 34, 35, 36, and 37.) Below 300 KIAS, the turning rate and radius of the aircraft improved, as did its acceleration performance. A loss of sustained maximum performance maneuvering capability imparts offset these advantages. Above 350 KIAS, other considerations arose and amplified upon as this discussion progresses.

Straight, level, and unloaded acceleration comparisons at both military and maximum power favored the USAF tactical fighters at high-calibrated airspeeds. The Fresco C had an excellent acceleration potential near 250 KIAS. However, and frequent mention occurred in the summaries of no significant advantage befalling USAF fighters until airspeeds approached or, in the case of the F-100, exceeded 300 KIAS.

On TAC flight 1, initiating a military power acceleration comparison with an F-4 at 250 KIAS enabled the Fresco C to exceed the speeding up of the F-4E before 260 KIAS. During the evaluation of certain offensive and defensive maneuvers, the Fresco C at low calibrated airspeeds displayed a consistent ability to out-accelerate the participating fighters.

The exploitation team conducted turn performance comparisons on numerous flights during this evaluation at speeds of 350 KIAS, 450 KIAS, and Mach .9. At 350 KIAS, the Fresco C enjoyed a turn superiority over the F-100, F-105, and F-4. It realized a parity with the F-4 decreasing airspeeds from that point increased the margin of superiority against the three aircraft mentioned. At 300 KIAS, the MiG-17F could out-turn the F-5.

Increasing the airspeeds from 350 KIAS, however, presented a somewhat different situation. At about Mach .85 or 450 KIAS, whichever was less, the unboosted rudder and elevator of the Fresco C encountered aerodynamic forces opposing the deflection of the control surfaces. The exploitation team noted this tendency on the initial TAC flight. They considered this a characteristic of the aircraft throughout the evaluation; consequently, referring to the turn comparisons conducted at 450 KIAS and Mach .9 not altogether impressive.

Another factor then presented itself when the airspeed lost in the execution of a maximum performance turn. Although at high-calibrated airspeeds, the F-105, F-100, and F-5 completed a 180-degree turn before the Fresco C did so. The F-4 matched turn rates with the Fresco as the MiG-17 invariably lost less airspeed in the turn.

A canopy mounted periscope assembly enhanced visibility from the cockpit in the rear hemisphere over the pilot's head, affording visual lookout of approximately 20 degrees to either side of the tail, and 10 degrees up or down. The overall visibility was not good.

The seat sat low in the cockpit, and even when topped, the pilot had trouble seeing over the nose and sides. Additionally, the periscope assembly itself obstructed overhead visibility and inconvenienced the pilot.

The Fresco C was extremely small and very difficult to see, particularly if one was unaware of its presence. It did not emit a smoke trail of any significance whatever, which complicates the visual detection problem even more. Noting the short (1–3 NM) visual acquisition ranges and positive ID ranges recorded by on flights 33 34, and 35 proved interesting. Those ranges contrasted with the ranges at which the Fresco C pilot visually acquired the F-4s (10–15 nautical miles).

The Fresco C had undampened flight controls. The aircraft became unstable in all axes at 375 KIAS, and higher and directional control was poor after that. This characteristic was encountered in TAC flight 1 and noted throughout the test became significant when the aircraft attempted to track an airborne or ground target at that speed. Further, yaw control of the plane became difficult when disturbing the airflow over the flight controls. Consequently, the aircraft was unstable in yaw at any speed in turbulent air

The Fresco C flew in a clean configuration on all the 57 TAC flights flown, some configuration yielding flight durations ranging from 25 to 45 minutes, depending on the type flight flown. Obviously, a flight requiring frequent resort to the utilization of maximum power enabled less time aloft than did a flight requiring lesser power settings. Sustained low altitude maximum performance maneuvering by the Fresco C exhausted the available internal fuel in 20–25 minutes. It was, therefore; apparently better to use the aircraft in a point defense role, wherein the fuel consideration was not prohibitive.

The maximum speed of the Fresco C at maximum power in level flight, per FTD-CS-O9–5–67, was about Mach.96 in level flight. For effective employment as a weapons system, the Fresco C should not fly at high-calibrated airspeeds because of the previously mentioned high stick forces affecting the maneuverability of the aircraft

The speed brakes of the Fresco C operated very efficiently. Evaluated in comparison to those of the participating aircraft on flights 1, 8, 22, and 50, they varied by comparison. However, in all cases, they remained more effective than those of the other participants did. This fact remained true, regardless of the power settings used with comparable power settings maintained throughout in both aircraft

The Fresco C flew and evaluated against a towed TDU-10B dart on six different occasions besides four flights dedicated to the evaluation of the Fresco C in an air-to-ground strafing role. The exploitation team obtained good results under ideal conditions (low attack speed, smooth air, and low g). However, in less than ideal conditions (higher attack speed, turbulent air, higher g), the results remained less favorable. Numerous considerations detracted from the effectiveness of the armament. As previously discussed, the exploitation team found the Fresco C unstable in turbulent air and at airspeeds greater than 375 KIAS.

The exploitation found the MiG-17 to have low firing rates. The two 23 mm cannons fired at the rate of 900 rounds per minute, and the 37-mm cannon fired at 400 rounds per minute rate. With all three guns simultaneously firing, the total cumulative firing rate equated to 2,200 rounds per minute. This was a low rate when compared to the 6,000 rounds/minute rate of the Gatling gun on the F-4 and F-105.

The exploitation found the MiG-17 also to have a low muzzle velocity. The muzzle velocity of both projectile sizes was 2,250 feet per second. By comparison, the muzzle velocity of USAF 20 mm projectiles was 3.280 feet per second.

To meet the excessive tracking requirement, the optical sight provided a proper lead computation after two to three-seconds of steady-state tracking. The dynamics of an aerial engagement could yield the requirement prohibitive.

The low muzzle velocities of the projectiles and their weights precipitate a demand for a large amount of lead as compared to the USAF 20 mm weapon. In a maximum performance maneuvering engagement, the extent of lead required could exceed the ability of the aircraft to obtain it.

The project officer used two sources of Fresco C performance data in conducting this test and found them slightly divergent. He used as sources, FTD-CS-O9–5–67, Fresco (MiG-17) Weapon System, 18 July 1967 (S), and APGC-TR-66–4, dated March 1966 (S). The former more accurately described the performance characteristics of the test bed aircraft after the AFFTC completed a comprehensive flight test evaluation of the aircraft

The aircraft flown on this assessment proved highly reliable and maintainable. Twenty-three major discrepancies occurred on 224 flights between 1 February 1969, and 14 March 1969 and in that period with only two flights lost to maintenance. These figures appeared even more impressive considering the aircraft frequently flew four flights in a single day.

USN Evaluation
Performance Comparison

AREA OF COMPARISON	F-4	F-8	A-4	A-6	A-7
ACCELERATION					
MIL	Sig. Superior	Superior	Superior	Superior	Inferior
MAX	Sig. Superior	Sig. Superior	Inferior	Inferior	Inferior
DYN. PRESSURE LIMIT	Sig. Superior	Sig. Superior	Comparable	Superior	Comparable
DECELERATION					
With Power	Inferior	Inferior	Inferior	Inferior	Superior
Without Power	Comparable	Inferior	Inferior	Inferior	Superior
ROLL RATE	Superior	Superior	Superior	Superior	Superior
TURN RADIUS					
Above 450 KIAS	Superior	Superior	Superior	Superior	Superior
Below 450 KIAS	Inferior	Inferior	Inferior	Inferior	Inferior
TURN RATE					
Above 450 KIAS	Superior	Superior	Superior	Superior	Superior
Below 450 KIAS	Inferior	Inferior	Inferior	Inferior	Inferior
ZOOM					
MIL	Sig. Superior	Superior	Comparable	Comparable	Comparable
MAX	Sig. Superior	Sig. Superior	Inferior	Inferior	Inferior

Chapter 10 - Air Force Carry-on MiG Projects in Nevada

Air Force Assumes CIA Operations at Area 51

Project HAVE PRIVILEGE

On 25 November 1969, a Cambodian Khmer Air Force pilot defected to South Vietnam in the Chinese copy of the MiG-17F, the Shenyang J-5.

To exploit this MiG, the AFSC/FTD tapped Col Wendell Shawler, the USAF pilot, who flew the MiG-17F in HAVE FERRY and HAVE DRILL. Colonel Shawler went to go to South Vietnam and made several evaluation flights of the J-5 to establish it having the same flight characteristics as the MiG-17. This short program of just five flights from Phu Cat Air Base in South Vietnam was code-named HAVE PRIVILEGE.

Resulting from these four top-secret exploitation programs, both the USAF and Navy changed their fighter tactics and trained their pilots again to exact as much capability and performance out of the aircraft as possible to win the dogfight. In 1989, a Pentagon official finally confirmed the 1981 combat of two US Navy F-14 Tomcats versus two Libyan Sukhoi Su-17 fighters. These occurred over the Gulf of Sidra used tactics developed out of mock combat testing with US-operating Soviet fighters.

At least one more MiG-21 captured in Vietnam and a MiG-25 flown by defector Soviet Pilot Viknotsor Belenko arrived at Area 51 where they flew before returned to the source. In 1973, the exploitation continued with Project HAVE IDEA, which took over from the older HAVE DOUGHNUT, HAVE FERRY, and HAVE DRILL projects. In July 1975, the 4477th TEF ("Red Eagles") formed at Nellis AFB, In December 1977, the 6513th Test Squadron ("Red Hats") formed at Edwards AFB to continue testing this foreign craft. Some aggressor training occurred where the units went head to head against USAF fighters in mock dogfights now to find out and exploit possible weaknesses.

DEPUTY DIRECTOR OF CENTRAL INTELLIGENCE

WASHINGTON, D.C. 20505

26 August 1976

C/S has noted

27 AUG 76

General David C. Jones
Chief of Staff
United States Air Force

Dear General Jones:

You will recall that in an exchange of messages between you and Mr. Colby in February and March 1975 (OMAR 8020 and PADRE 5406) you expressed the desire to have CIA continue to manage Area 51 at the Nevada Test Site. Mr. Colby agreed to retain management of the Area "as long as community intelligence requirements and budgetary constraints permit." Subsequently, the Senate and House Appropriations Committees deleted Area 51 operating funds from the FY-76 CIA appropriation, but in accordance with USAF desires, recommended that CIA continue to manage the Area with the operating funds being provided by the Air Force. That recommendation was implemented, and we have been operating under the arrangement for a little over a year.

In a recently completed review of the operation of the Area, including the projects currently underway there and those planned for the future, it was concluded that management of the Area by this Agency is no longer warranted. Essentially all of the activity at the Area is in support of DOD projects, the Air Force's HAVE GLIB project being by far the largest at the present time. Another large Air Force project and a large Army project are projected for the near future. CIA's use of the Area for its own projects is minimal, and we see no significant increase in our requirements for the future. Although it is true that the DOD projects are intelligence related and do require a covert base, I am convinced that the Air Force could manage the Area and maintain its covert posture. In addition, it would appear to be better management practice to have the funding organization responsible for the operational management of the Base.

This issue was brought up before the Committee on Foreign Intelligence (CFI) as a key topic in the review of the Agency's FY-1978 program. The CFI approved the recommendation that management of Area 51 be transferred from CIA to the Air Force by FY-78. We therefore plan to begin a gradual phase-out that will be complete by the end of FY-77. We hope that by beginning now we can effect a smooth transition by that time with little or no effect on tenant projects.

**Released in
October 2013**

Sincerely,

E. H. Knoche

18 72:76

E-2 IMPDET
CL BY 034944

SECRET

Project HAVE PAD

MiG-23 flying over Nevada

In 1978, the AFFTC, TAC, and the US Navy exploited the Soviet MiG-23 Flogger at Area 51 in a project code-named HAVE PAD. The primary objective of HAVE PAD determined the performance and flying qualities of the MiG-23MS with a secondary objective of conducting a tactical evaluation to determine its combat effectiveness against US aircraft

The Flogger MiG-23MS, powered by an R29–300 power plant, was an export fighter-interceptor capable of a dual role in air-to-air intercept or ground attack. It was a versatile aircraft because of its inherent interceptor design.

It had variable geometry wings, AI radar, and multiple weapon stations, navigation, and communications equipment (RSBN and data link). It flew at a high, top speed of Mach 2.35 and had a low landing speed of 140 knots and excellent acceleration characteristics. Also, it had a maximum endurance capability of 4.3 hours cruising at 10 km altitude at Mach .65 using 3/800-liter external fuel tanks. Its maximum operational range was 1080 nautical miles without external fuel and 1565 NM with 3 x 800-liter fuel tanks.

Access to the MiG-23 weapon system made its exploitation important to the US Air and Navy who believed the MiG-23 an excellent dogfighter. Evaluation of the AI radar was also necessary as the Soviets now used it in the late model MiG-21 Fishbed planes, making them a quantitative and qualitative threat. The subsystems of the 1974 production MiG-23 represented the heart of nearly every Soviet late model domestic and exporter fighter aircraft flying at the time. The technology benchmark recorded in this plane coupled with the earlier MiG-25 exploitation provided a unique database upon which to judge current and future Soviet capabilities better. Its rugged tricycle landing gear allowed operation on both concrete and sod airfields.

The exploitation team found that despite its small size, excellent engine response, nearly jam-proof fire control radar system, and gun/missile weapons load, the Flogger was not an exceptionally capable dogfighter. They found its turning performance unimpressive, its cockpit visibility poor, the engine smoked, high flight control forces and the variable wing sweep system neither optimized nor automated. They found the pilot head-down aids not optimized for the aerial combat arena.

The exploitation of the HAVE PAD plane's avionics systems afforded an unparalleled opportunity to examine, test, and photograph current Soviet avionics equipment over an extended period yielding some very significant results.

The ARK-15M radio compass was the first military avionics system to have integrated microcircuits in two-sided, printed circuit board, plug-in cards, and digital processing. Testing the JAY BIRD AI radar confirmed the operating frequency range of 12,730 - 13,220 MHz and the search/track ranges of 30/20 km.

The JAY BIRD was very like the I-band portion of the FOXFIRE in configuration, physical appearance, and operation. Peak power was about 170 kW, higher than the 100 kW previously estimated. The JAY BIRD contained an impressive collection of ECCM features, making it highly resistant to conventional jamming techniques. The testing of the rate gyros of the SAU-23 automatic control system in the laboratory at Groom Lake made it the first exploitation of Soviet aircraft gyros.

The armament complement of the Flogger was a good mix for a fighter-interceptor with the internal gun, ATOLL missiles, and a variety of air-to-surface weapons. Unfortunately, no ATOLL missiles came with the plane, denying the exploitation team an opportunity to exploit the semi-active (SA) missile, leaving a gap in knowledge of a significant threat item. However, the exploitation team found weak points in the Flogger weapons area. These included the inability to launch single bombs or rockets, poor switching requirements, no head-up display, and no off-axis missile lock-on capability.

The R29–300, used in the MiG-23MS, was the latest known operational S.K. Tumanskiy power plant. Although similar in design philosophy to previously exploited Tumanskiy engines, i.e., R11, R13, and R15, those power plant reflected a significant improvement in airflow handling capacity, thrust/weight, turbine inlet temperature, engine control logic and acceleration characteristics. Additionally, the engine used a unique variable turbine cooling scheme, which permitted essentially turning off the cool air when not required, thus improving subsonic cruise specific fuel consumption. The engine operated due to the flight test program for approximately 87 hours with no major propulsion problems or pilot complaints.

This engine reflected a continuation of Soviet air-breathing propulsion design philosophy. The major driver was on engine cycle selection. Its design called for low cost, simplicity, and operational capabilities, placing little emphasis on the time between overhaul, manufacturing man-hours, or on the range. In the structures area, the MiG-12 employed essentially the same materials and manufacturing techniques found on earlier Soviet aircraft

The Flogger used an RSBN-6s short-range navigation and landing system airborne orthodromic navigation system in conjunction with the ground-based transmitter to perform tactical navigation and instrument landing. The pilot could select a preprogrammed destination — one of four ground stations or one of three route waypoints (points defined on the ground stations) — or home base.

Its ALMAZ-23 (JAY BIRD) airborne intercept (AI) radar provided target search, acquisition, and grand and air-to-air missile (AAM) target illumination. JAY BIRD operated in the J-band, which provided a sharper beam, better angular resolution, and better control over side lobes. The radar maintained automatic tracking of the target under most target jamming conditions. If noise jamming denied range tracking, the radar could still track-on-jam (TOJ) or derive angle data by tracking the jamming source. Range data supplied by data link provided the remaining information necessary for a missile launch. The SIRENA-3M radar warning system provided passive warning of incoming radar signals to the pilot by a lighted display and a tone in the headphones.

A rather significant finding of HAVE PAD exploitation was the use of V95 aluminum alloy for the primary wing skins, an alloy that had caused the US serious metal problems through the 1960s.

The well laid out cockpit of the Flogger had many human factors apparent. It was also an obvious Mikoyan design very like the later Fishbed MiG-21 and FOXBAT MiG-25.

The canopy of the Flogger was a pneumatically operated clamshell that hinged aft for access. The emergency air operated jettison system on the Flogger was independent with a type KM-1M ejection seat system designed to operate from 70 - 540 KIAS at altitudes from ground level to over 18,000 m.

The plane used a catapult, rocket-powered seat using four pyrotechnic devices and numerous precooked springs for sequencing and actuation. The pilot initiated ejection by gripping one or both ejection handles, pushing the locking lever(s) to unlock a ball lock securing the handle to the seat pan and pulling the handles(s) upward.

During the initial vertical movement, a lanyard-actuated spring ensnared the pilot's legs to prevent flailing. After the seat cleared the aircraft, a three-chute stabilization, speed-retardation, and personnel-descent system sequenced the chute deployment and released per the ejection speed and altitude.

The Soviet VKK-4 and VKK-6 partial-pressure suits appeared like the USAF model MC-4 suit. A capstan system provided anti-G protection extending along both legs from approximately the bottom rib down to the ankle region. The Flogger pilot used one of two types of standard flying helmets, depending on the mission. The pilot wore a crash helmet for subsonic flights at low to medium altitudes and a pressure helmet for high-altitude and high subsonic flights.

The NWC conducted the ground-based IR measurements of the Flogger. However, other commitments negated the use of some of the EG&G special projects' radar test facilities, this limiting this exploitation to the recorded IR spectral data recorded by contractor General Dynamics during the ground-to-air tests.

One aircraft sortie was devoted to radar signature measurements, obtaining I and G band aspect measurements of the HAVE PAD vehicle.

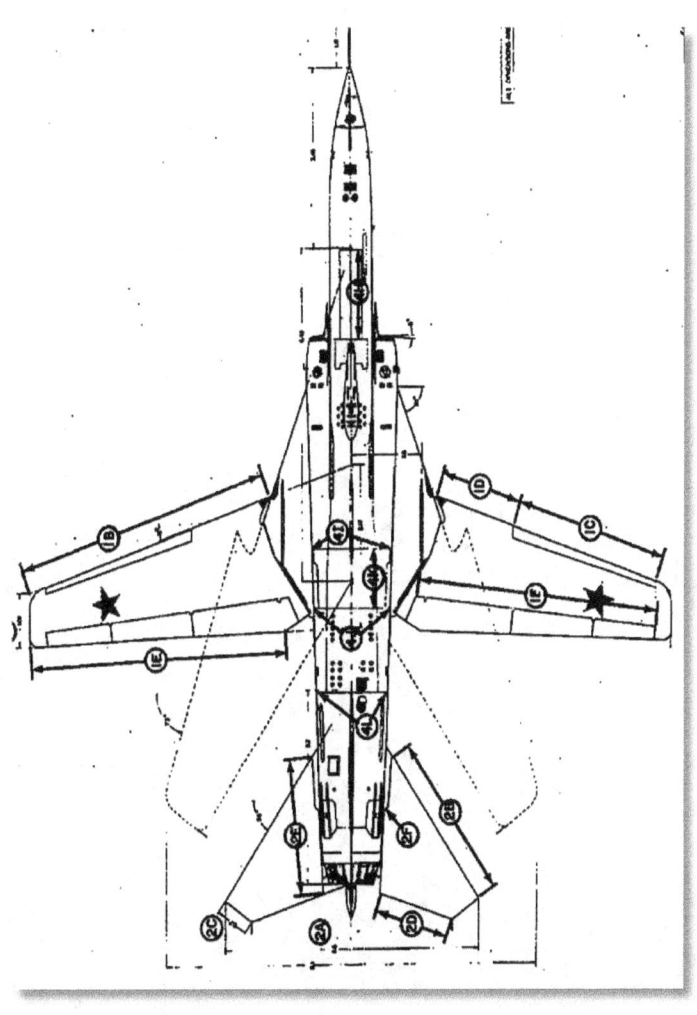

Epilogue

Even today the US and its ally aviators routinely test their skills against their adversary's latest plane available. Pilots, including the famous astronaut Lt. Gen Thomas Stafford of Weatherford, Oklahoma, continued to fly the Nevada skies long after the HAVE DOUGHNUT, HAVE Drill, and HAVE FERRY programs.

Published speculation held that at one time or another since the first MiG arrived at Groom Lake in 1968, the US has flown in the Nevada desert every warplane in the Soviet arsenal, including at least two MiG-23 Floggers confirmed by actual sightings. Black programs began by HAVE DOUGHNUT gaining intelligence by flying additional Soviet-built fighters in the US reportedly peaked in the late 1980s and ended in the 1990s. The training of US pilots in the ways of their adversaries did not cease.

On April 26, 1984, Lt Gen Robert M. Bond, who had flown the `HAVE' MiG 15 years earlier-died in the crash of a MiG-23 Flogger. The crash occurred on the Nellis range, one of the few occasions on which the USAF could not cover up post-1969 flights of the Soviet MiGs. Sadly, the general considered this his farewell flight to mark the end of a fine career.

The high-profile Bond mishap was a MiG accident the US Air Force could ill afford. Politics, rather than real security concerns accounted for most of the secrecy. Obviously, the Soviets knew what was going on. The participants kept the tests of this aircraft secret, primarily to protect the identity (or identities) of the country (countries) providing the MiGs flown in Nevada.

Ironically, the latest in Soviet aircraft flooded the world following the end of the Cold War. Virtually every aviation museum in the United States had on display the infamous MiG 21 plus most of the planes built afterward. Nellis AFB in Nevada currently hosts a threat museum affectingly referred to as the petting zoo because of the museum's popularly among civilian tour groups wanting to have their photos taken in the cockpit of a MiG 29.

NAS Fallon displays one of the finest collections of MiG planes that one can imagine. Almost every aviation museum in the United States now has Soviet MiGs on display. The American appetite for displaying the Soviet MiG is a testimony to the lasting role the Soviet planes in American military aviation.

This current appetite to learn more about these mysterious planes started with the CIA's special projects team of inspired specialists. With time on their hands between missions, they had the curiosity, the knowledge, and equipment to experiment with learning the radar cross-section signatures of just about any plane available. These included the MiGs.

The twenty-four radar intercept sorties accomplished by the F-4 to evaluate the MiG-21 gave them the opportunity to compare radar signature characteristics. The smallest radar cross-section occurred from the head-on aspect, an average detection range was 20 nautical miles,

Lock-on averaged 15 nautical miles from the head-on aspect. From a tail-on aspect, ranges increased to 25 nautical miles and 17 NM respectively. From abeam, or 90-degree aspect, the range for acquisition and lock-on rose to 35 and 28 nautical miles, respectively. They determined target altitude a consideration only because ground clutter at lower altitude complicated the radar target recognition problem.

Comparison of APQ-109 (F-4D) radar detection ranges and those of the APQ-120 (F-4E) revealed the APQ-109 acquiring the MiG-21 at a 510 percent greater range. The source of such a small deviation over a limited sampling was hard to define. One had to recognize the increased beam width of the APQ-120 decreasing the amount of power concentrated on the target. This required having a redesigned radar reflector compatible with the internal gun of the F-4E. Thus, they expected a slight degradation in range performance.

Neither the F-4D nor F-4E came currently equipped with a suitable low altitude or look down capability against airborne targets. It, therefore, became evident the vulnerability of both aircraft (and the

F-4C as well) from a radar detection standpoint.

Project AQUILINE

In 1965, two engineers from the Office of Research and Development issued a request for proposals for a small, unmanned drone. Douglas Aircraft was the only company to respond.

Project AQUILINE was a stealthy propeller-driven, low altitude, anhedral-tailed UAV called Aquiline and designed for low-level electronic surveillance of the Chinese nuclear program. Aquiline was designed by McDonnell Douglas in the late 1960s and advanced to flight testing, but never saw operational use due to reliability problems. A data link from a high-flying U-2 controlled the aircraft.

In the early 1960s, there were many problems in obtaining coverage of hostile territory. The U-2 was too vulnerable to Soviet surface-to-air missiles, as had been demonstrated by losses over the Soviet Union, Cuba, and the People's Republic of China. The OXCART was still under development and even when completed might prove vulnerable to Soviet radars and missiles. Although safe from interception, the newly developed photo satellites could not provide coverage of a desired target on short notice.

Because several of the intelligence community's primary targets such as Cuba and the new Soviet radar installation at Tallinn (Estonia) were not located deep in hostile territory, CIA scientists and engineers began to consider the possibility of using small, unmanned aircraft for aerial reconnaissance. They believed that recent advances in the miniaturization of electronic technology would make possible the development of a reconnaissance vehicle with a very-low-radar cross-section and small visual and acoustical signatures. Such a vehicle could reconnoiter an area of interest without the hostile country realizing that it had been overflown.

Aquiline was essentially a powered glider with an 8.5-foot wingspan. The aircraft weighed only 105 pounds. Aquiline's tail-mounted 3.5 horsepower chainsaw engine drove a two-bladed pusher propeller. Powered by a small 3.5-horse-power two-cycle engine originally developed by the McCullough Corporation for chainsaws. The vehicle was a six-foot long plane that looked like an eagle or buzzard when it was in the air. It was designed to fly at very low levels along communications lines and intercept their messages. It also had a small television in the nose as an aid to navigation and to photograph targets of opportunity. This small vehicle was launched from inclined rails and was recovered in a large net strung between two poles.

John Henry "Hank" Meierdierck had first come to Area 51 in 1955 as an instructor pilot to train the CIA pilots in the U-2. Meierdierck had retired from the Air Force and at the urging of Gen Jack Ledford, accepted a GS-14 assignment with the CIA's OSA, Office of Special Activities where he maintained operational control of the three Agency U-2 detachments deployed overseas. Following the shoot-down of Gary Powers, he became a GS-15 and the operations chief over the A-12 and YF-12 projects occurring at Area 51.

For the next 8 years, he would spend most of his time commuting to Area 51, monitoring the program, and developing operational tactics, arranging the spending of the funds and the development of payload packages. He also was the OSA budget officer and through this had a say in all the operations and projects, since he had to defend each line item to higher headquarters. He was assigned to Plans and Programs and had control of the budget besides being the contact man for operations with Area 51 and overseeing the U-2 operations and the development of the SR 71.

At the end of the CIA's A-12 OXCART program, the CIA asked Meierdierck if he would like to return to Area 51 as the commander of a new airplane project named Aquiline. He, of course, jumped at the chance. He and his wife, Millie, packed up the household goods and headed west towing a horse trailer with their daughter Vicki's pony in it.

Project Aquiline was a project to develop an RPV [Remotely Piloted Vehicle] The contractor was Mc Donald Douglas. He had an admin staff headed by a CIA professional, Mr. Andy Frisina and three pilots that were to be trained to fly this "BIRD," and a maintenance officer. His small but knowledgeable group was stationed at Area 51 and commuted about once a week, depending on the flying schedule.

The contractor had the responsibility for the development and test, and Meierdierck's group was to assume command as soon as Meierdierck declared the vehicle 'Combat Ready' Things did not start off too good with the contractor since they did not want him or the CIA looking over their shoulder. They tried to keep everything secret.

Progress was very slow because of and some crashes [reason unknown] and some lousy landings. One or more of the aircraft was always under repair, and eventually, three of the five AQUILINE prototypes were destroyed in testing. When it became budget time, the contractor predicted the amount of money needed since Meierdierck did not have the intimate knowledge of the development expenses. He had 11 million dollars for the following year, and he advised Macdonald of this fact and asked for the next years operating budget.

The contractor came back to him with a ridiculous $110 million forecast. Meierdierck returned to CIA headquarters where discussed it with the bosses, who suggested he give Macdonald Douglas two weeks to adjust the amount and then to come to Washington, DC to present their budget.

The company decided to back into the $110 million number rather than justify the true amounts needed. During their presentation, Meierdierck was forced to interrupt them many times to point out errors and outright lies. Upon the completion, the CIA management asked for his comments.

Meierdierck explained that they needed only the $11 million and could not possibly spend the larger amount. He also pointed out their exaggerations, padded costs for items and the brazen lies that they tried to force on the CIA. When asked what he thought the CIA should do, he stated, directly that since this company tried to charge $110 million for a $11 million job, that he could not trust them and that the CIA should cancel the program. The CIA agreed and canceled the program.

By 1970, the HAVE Drill program expanded where a few selected fleet F-4 crews got the chance to fight the MiG. The most important result of Project HAVE Drill was that no Navy pilot who flew in the project defeated the MiG-17 Fresco in the first engagement.

The HAVE Drill dogfights occurred by invitation only. The other pilots based at Nellis Air Force Base knew nothing about the US-operated MiG. To prevent any sightings, the Air Force closed the airspace above the Groom Lake portion of the Nellis Range. Aeronautical maps depicted the exercise area with red ink borders, the forbidden zone known as "Red Square."

USAF Colonel Gail Peck, a Vietnam veteran F-4 pilot dissatisfied with his service's fighter pilot training, devised the idea of a more realistic training program for the Air Force. After the war, he worked at the Department of Defense, where he heard about the HAVE DRILL and HAVE DOUGHNUT programs. He won the support of USAF General Hoyt S. Vandenberg, Jr., and launched "CONSTANT PEG," named after Vandenberg's call sign, "Constant," and Peck's spouse, Peg.

In the mid-1970s, the United States sold fighter aircraft to Indonesia and Egypt to replace the Soviet fighter planes, allowing these nations to transfer unneeded MiG-21 ultra-modern MiG-23s aircraft to the United States clandestinely for evaluation. The aircraft were acquired from other friendly nations such as Israel, Indonesia, Egypt, etc., who either purchased them for their own arsenals or captured them during wartime. We also acquired Soviet type radar and threat systems, missiles and other hardware.

Up to 25 of these Soviet aircraft made their way to Groom Lake. Pilots assigned to Detachment 1,

57th FWW at Nellis went to the facility for training as "Aggressor" pilots. From there, they received reassignment to the aggressor training units at Clark AB, Philippines, RAF Alconbury, England, and Nellis AFB. However, by the mid-1970s, the fleet of Soviet aircraft grew at Groom Lake, crowding the facilities. The Soviet Union knew of the MiGs at Area 51 and monitored their activities, so the MiG needed another clandestine home.

Photo courtesy the 4417th TES

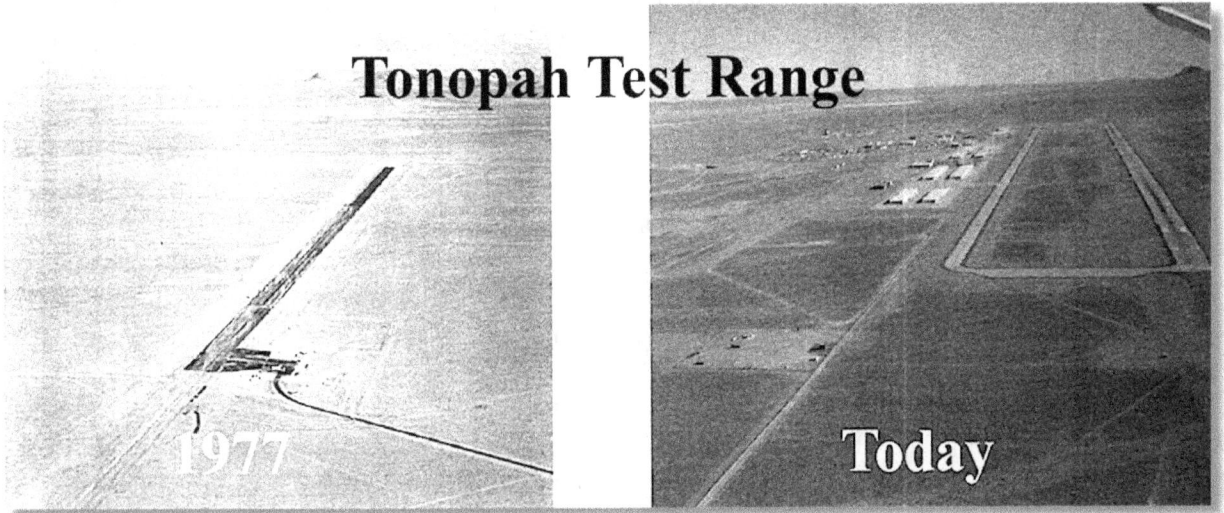

The several locations considered included Michael Army Airfield at the Dugway Proving Grounds in Utah and the Gila Bend Air Force Auxiliary Field on the Goldwater Range in Arizona. However, the Tonopah Test Range Airport was only 70 miles to the southeast of Groom Lake and was on the controlled AEC

Tonopah Test Range fitted the need for a new home. The AEC airport had the potential for improvement and expansion, with the only public land overlooking the flight test facility miles away.

Although not as hidden as Groom Lake, the airport was remote enough to operate the Soviet aircraft. The security surrounding the Tonopah Test Range was so effective that no one publicly reported the new flight test facility as an Air Force military airfield until 1985.

On 1 April 1977, Air Force reassigned the 4477th TEF, a Tactical Air Command unit to Tonopah. In December 1977, Edwards AFB formed the 6513th Test Squadron ("Red Hats"), a Air Force Systems Command, AFSC unit to perform technical evaluations of these aircraft. The 4477th would later share the TTR base with the 4450th Tactical Group flying the stealthy F-117 Nighhawk and its A-7D chase planes.

The product of Project CONSTANT PEG, the 4477th Test and Evaluation Squadron (4477 TES), based at a remote airfield at Tonopah Test Range. TTR, located in the desolate desert north of Las Vegas, became a squadron in the United States Air Force under the TAC.

There, they trained USAF pilots and weapon systems officers, and USN and USMC Naval Aviators and Naval Flight Officers to better fight the aircraft of the Soviet Union. Some 69 pilots, nicknamed Bandits, served in the squadron between 1979 and 1988, flying Soviet-built aircraft (MiG-21MF, MiG-23MS, MiG-23BN and Su-20) in mock-combats against USAF aircraft.

The author cannot reveal the actual number or types of aircraft involved, where they came from, or the complete history of the program. As revealed through declassification, the activities of the 4477th Test and Evaluation Squadron caused a fundamental change in the United States Air Force, Navy, and Marine Corps air combat tactics. They revitalized the near-forgotten art of dogfighting.

The squadron's knowledge gained from testing the aircraft the squadron flew reflected in the success of United States air operations during the Vietnam War by it supporting the Air Force's RED FLAG program and the United States Navy's TOPGUN School.

In May 1973, Project Have Idea formed to take over from the older HAVE DOUGHNUT and HAVE Drill projects and the project transferred from the Area 51 facility to the Tonopah Test Range Airport, Nevada. At Tonopah, testing of foreign technology aircraft continued and expanded throughout the 1970s and 1980s.

By the late 1970s, United States MiG operations underwent another change. In the end of the 1960s, the MiG-17 and MiG-21F remained the frontline aircraft. A decade later, later-model MiG-21s and new aircraft, such as the MiG-23 superseded them.

Fortunately, a new source of supply of Soviet aircraft became available, Egypt. In the mid-1970s, strained relations between Egypt and the Soviet Union caused Egypt to order out its Soviet advisers.

The Soviets had provided the Egyptian Air Force with MiG since the mid-1950s. Now, with their traditional source out of the picture, the Egyptians began looking west. They turned to United States companies for parts to support their late-model MiG-21s and MiG-23s. Very soon, they entered a deal where per one account,

Egyptian President Anwar Sadat gave the United States two MiG-23 fighter-bombers disassembled and shipped from Egypt to Edwards Air Force Base. They then transferred initially to Groom Lake for reassembly and study.

In 1987, the US Air Force bought 12 new Shenyang F-7Bs from China for use in the CONSTANT PEG program. At the same time, it retired the remaining MiG-21F-13 Fishbed planes acquired from Indonesia.

The United States operated MiG received special designations. There was the practical problem of what to call the aircraft to solve this; the US gave them numbers in the Century Series. The MiG-21s and Shenyang F-7Bs became the "YF-110" (the original designation for the USAF F-4C), while the MiG-23s became the "YF-113."

The focus of AFSC limited the use of the fighter as a tool with which to train the front line tactical fighter pilots. AFSC recruited its pilots from the AFFTC at Edwards Air Force Base, California. Most of them graduated from the Air Force Test Pilot School at either Edwards or the Naval Test Pilot School at NAS Patuxent River, Maryland. TAC selected its pilots primarily from the ranks of the Weapons School graduates at Nellis AFB.

The 4477th began as the 4477th Test and Evaluation Flight (4477 TEF) on 17 July 1979, four years after the Air Forces' first Red Flag Exercise and 11 years after the Navy's first Top Gun training exercise. The name later changed to the 4477th Test and Evaluation Squadron (4477 TES) in 1980. The 4477th began with three MiG: two MiG-17Fs and a MiG-21 lent by Israel, who had captured them from the Syrian Air Force and Iraqi Air Force. Later, it added MiG-21s from the Indonesian Air Force.

The Air Force collected the aircraft at the Department of Energy's Tonopah Test Range, where the squadron flew the MiG-17s until 1982. They mostly flew MiG-21s and MiG-23s.

Two pilots of the 4477th died flying the Soviet planes. The pilots had no manuals for the aircraft, although some tried to write one, nor was there a consistent supply of spare parts, which the squadron refurbished or manufactured at high cost.

On 23 August 1979, a pilot lost control of the squadron's MiG-17F, USAF serial 002, the HAVE FERRY MiG from Area 51. US Navy Lieutenant M. Hugh Brown, 31, of the US Navy's Test and Evaluation Squadron FOUR (VX-4), "Bandit 12," originally of Roanoke, Virginia, entered a spin while dogfighting a US Navy F-5. Brown recovered, however, entered a second irrecoverable spin too low to eject. The plane hit the ground at a steep angle near the Tonopah Test Range airfield boundary, killing the pilot instantly.

On 21 October 1982, USAF Captain Mark Postal crashed with a MiG-23.

On 26 April 1984, USAF Lieutenant General Robert M. "Bobby" Bond, then vice commander of AFSC, died attempting to eject after losing control of his MiG-23 while supersonic. A few hours after the crash, Headquarters, AFSC, at Andrews Air Force Base, Maryland issued a brief statement: "Lt. Gen Robert M. Bond, Vice Commander, AFSC, killed today in an accident while flying in an Air Force specially modified test aircraft." Three-star generals do not generally fly test flights, so Bond's death attracted press interest. The Air Force is refusing to identify the type of plane also raised questions. Early reports claimed him flying "a super-secret Stealth fighter prototype." The death of a three-star general led the Air Force to reveal that it was flying Soviet aircraft. Named in his honor is a Boulevard cutoff between Eglin Air Force Base (main base) and Hurlburt Field in Florida.

Much like Area 51 at the time, the 4477th Test and Evaluation Squadron facilities at TTR offered Spartan, mostly doublewide trailers with the roofs weighed down by tires to prevent them blowing off in high desert winds.

The 4477th TEF was a full-fledged squadron, whose operation developed realistic combat training operations featuring adversary tactics. These were dissimilar air combat training and electronic warfare. Additional MiGs acquired over the years grew to approximately two dozen MiG aircraft, including ultra-modern MiG-23s (designated YF-113s) from various sources that remain classified today.

These include three Cuban pilots who brought their MiG to Florida, and some Chinese made MiG bought outright from China via the front company Combat Core Certification Professionals Company (CCCP). Three Syrians flew their MiG-23 and MiG-29s to Turkey in 1988 followed by Soviet Captain Alexander Zuyev in 1989.

The Red Eagles contended with un-maintainable planes with no spare parts or technical manuals available. Ground crews resorted to reverse engineering aircraft components and manufacturing them from raw materials. Two US pilots lost their lives in catastrophic crashes. However, they, along with the Red Hats at Area 51, did learn the strengths and weakness of the Warsaw Pact's fast movers. The information proved invaluable to NATO fliers in any East/West engagement. American pilots proved this in air-to-air combat every time they encountered these Soviet Cold War era aircraft in Iraq, Libya, Angola, and elsewhere.

As remembered by Col Gail Peck, historian, TAC created the 4477th Test and Evaluation Flight (TEF) to host the training program chartered by M/Gen Hoyt S "Sandy" Vandenberg in s program called Project CONSTANT PEG. CONSTANT was the "call sign" of M/Gen Vandenberg. PEG was the spouse of Major Gaillard (Gail) Peck, who initiated the program while working for M/Gen Vandenberg at the Pentagon.

The flight of CONSTANT PEG trained USAF, US Navy, and US Marine Corps combat fighter aircrews on the best ways to fight and win when encountering MiG aircraft in aerial combat. The 4477th TEF "stood up" at Nellis AFB, NV on 1 April 1977 under the command of Lt Col Glenn Frick. It transitioned to the flight base at the Tonopah Test Range airfield under the command of Lt Col Gaillard Peck in July 1979. The unit was terminated operations at Tonopah in March 1988 under the command of Lt Col Michael Scott. During these intervening years, Lt Cols Earl Henderson, Tom Gibbs, George Gunning, Phil White, and Jack Manclark also commanded the unit.

Photo courtesy 4417th TES

Photo courtesy of the 4417th TES

During the active MiG operations, the 4477 TEF name changed to the 4477 Test and Evaluation Squadron (TES). Shortly after the termination of flight operations the squadron "stood down," the Air Force salvaged and redistributed the equipment to other users and reassigned all personnel.

During the period of operations, the 4477th TEF/TES flew over 15,000 MiG sorties. The 4477th trained almost 6,000 Air Force, Navy, and Marine Corps aircrews to fight the MiG-17, MiG-21, and MiG-23 and win.

Lt Col Glenn Frick hired SMSgt Bobby Ellis as the Chief of Maintenance. Bobby's task included hiring necessary personnel and supervising the management and restoration of the aircraft made available to the 4477th TEF.

Glenn Frick and his operations officer, then-Maj Ron Iverson (Lt Gen Ret), hired the pilots to fly the aircraft

Some MiG flying was ongoing under cover of test operations. The birth of CONSTANT PEG marked the end of the "test charade." It signaled the beginning of MiG operations in earnest. It was wholly dedicated to training American airmen in how to engage and beat the MiG aircraft in close-in aerial combat (dog-fighting).

Gail Peck took command from Glenn Frick on 1 Oct 1978. He embarked upon the task of supervising the completion of all aspects of "standing up" the 4477th TEF, including the completion of the airfield, the restoration of the MiG aircraft and the initiation of flight operations.

When operations began at Tonopah Test Range in July 1979, the manpower consisted of 29 officers and enlisted personnel including three pilots US Navy aviators.

Throughout the period, the 4477th was a joint operation with the USAF and US Navy and Marine Corps pilots assigned. Initially, one civilian secretary supported the operations at her workstation at Nellis AFB, NV.

The initial equipage at Tonopah in July 1979 included eight aircraft, including two MiG-17s, NATO Fresco C planes, and six MiG-21 NATO Fishbed C/E aircraft. A Kenworth tractor truck with an 18-wheel trailer initially served as the only vehicle assigned to operations.

Through the good work of Mr. Matt Foley (RIP January 2013 at age 92), an AF intelligence officer, the force structure of RED EAGLE aircraft grew to reach a maximum of 27 flyable MiG airframes in 1985. Not all remained flyable at the same time, as the maintainers moved certain critical parts from one flyable aircraft to make another aircraft flyable. Five T-38s used for MiG "chase" augmented the fleet for new pilots flying the single-seat MiG for the first time. The squadron lacked having two-seater MiGs or flight simulators.

Initially, the RED EAGLE MiG pilots flew mission aircraft at Nellis such as Aggressor F-5Es or Weapons School F-4Es, F-16s and F-15s. RED EAGLE pilots eventually evolved into logging time and taking check rides for their proficiency requirements in the assigned T-38s. As an added benefit, the T-38s enabled the unit to fly VIP orientations and provide incentive flights for the people assigned to the TTR (both MiG and F-117 personnel).

The unit also had a small fleet of civilian light twin aircraft (Cessna 404s later replaced by Mitsubishi

160

MU-2s) used for transportation, mostly between Nellis and TTR. RED EAGLE pilots flew the light twins. The Air Force eventually phased out the light twins for AF C-12s flown by non-project military pilots.

The MiG-17s retired in 1981 after an engine problem prompted Capt Mark Postai to crash land in the desert. He stated after escaping uninjured that it was a very rough ride and something his not likely doing again.

The drawdown of the MiG-17 fleet corresponded with the introduction of the MiG-23 Flogger and the acquisition of additional MiG-21s. At the peak, the 4477th had 17 MiG-21s and 10 MiG-23s. An increased number of personnel assigned along with the acquisition of, and in some cases, the manufacture of additional surface vehicles accompanied the buildup of the MiG fleet. The RED EAGLES acquired these from the Defense Reutilization Management Offices (DRMO) or salvage yards throughout the West.

Three RED EAGLES died during the MiG program. LT (USN) Hugh Brown perished in the crash of the HAVE FERRY MiG-17 and Capt Mark Postai in the crash of a MiG-23. TSgt Rey Hernandez, a fuels specialist, passed away because of injuries from an industrial incident relating to the repair of a fuel cell in a RED EAGLE T-38 plane.

Eventually, the cost of keeping the MiG in the air caught up to the program. By the late 1980s, the decline of communist rule in the Soviet Union rendered CONSTANT PEG anachronistic. The Air Force ended the program in March 1988. However, the Air Force did not deactivate the squadron until July 1990.

During the 2006 declassification of the CONSTANT PEG program, the USAF held a series of press conferences about the former top-secret US MiG. The US MiG flew more than 15,000 sorties, and nearly 7,000 aircrews flew in training against dissimilar aggressors in the Nevada desert between 1980 and the end of the program in 1988 when the Air Force deactivated after a high accident rate.

Interviews with many of the pilots, maintainers, and engineers revealed how these programs provided valuable information that helped improve the survivability of US pilots in combat. Although crude by western standards, the Soviet airplanes proved robust and highly reliable. Through a technology exploitation program working closely with the "Foreign Technologies" department of Pentagon. that grew over several decades, US pilots were able discover the airplanes' weaknesses in combat situations.

The Red Hats

In December 1977, the 6513th Test Squadron ("Red Hats") formed at Edwards AFB. They formed to perform technical evaluations of the Soviet MiG aircraft at Area 51 and later flown by the Red Eagles at the Tonopah Test Range in Nevada.

Testing of foreign technology aircraft began at Area 51 in 1968 had continued and expanded throughout the 1970s and 1980s. The growth in foreign aircraft included an additional MiG-17, MiG-21, MiG-23, Su-7B, Su-22, and other aircraft undergoing intensive evaluations.

The Red Hats inactivated in October 1992 and reactivated immediately as the 413th Flight Test Squadron, providing test and evaluation capability for electronic warfare (EW) systems.

Project HAVE GLIB

Area 51 foreign materiel evaluation program HAVE GLIB is believed to have involved testing Soviet tracking and missile control radar systems. A complex of actual and replica Soviet-type threat systems began to grow around "Slater Lake" (the pond, which was named after the former Roadrunners commander), a mile northwest of the main base. They arranged them to simulate a Soviet-style air defense complex. The Air Force began funding improvements to Area 51 in 1977 under Project SCORE

EVENT. It is also believed that HAVE GLIB was primarily the follow-on effort to the HAVE DOUGHNUT and HAVE DRILL/FERRY evaluations of MiG-21 and MiG-17 fighters. Nonetheless, Defections with MiG planes continued

Over the years, increasingly aircraft added to the collection. On September 6, 1976, Lieutenant Viknotsor Belenko defected with his MiG-25 to Hakodate, Japan. The US interrogated and debriefed him for five months after his defection, and employed him as a consultant for several years after that. The MiG disassembled and examined, returned to the USSR in thirty crates. Belenko brought with him the pilot's manual for the Foxbat.

He expected to assist American pilots in evaluating and testing the aircraft. However, the Japanese government only allowed the US to examine the plane and do ground tests of the radar and engines.

In July 1988, two Syrian pilots defected with their MiG-29s to Turkey followed by another Syrian pilot in April same year who landed his MiG-23ML in Turkey.

In October 1989, Syrian pilot Abdel Bassem landed his MiG-23ML in was Israel.

The USAF continues a Foreign Materiel Acquisition/Exploitation program, although the extent of acquisitions and operations of that program is most likely classified and unavailable. In March 1991, in the aftermath of the 1991 Gulf War, an exploitation team from the Joint Captured Materiel Exploitation Center arrived at Jalibah Southeast Air Base in Iraq. They returned with a MiG-29 nose, providing Air Force intelligence personnel with a Slot Back I radar and the Fulcrum's infrared search and tracking system.

Later in the decade, Air Force intelligence personnel acquired complete versions of the MiG-29, the result of spending money rather than fighting a war. In October 1997, the US bought 21 fighter aircraft from the Republic of Moldova—including the MiG-29UB.

According to the National Air and Space Intelligence Center, after "undergoing years of study" and employing "all the [center's foreign materiel exploitation] resources," the MiG-29UB was displayed in front of NASIC headquarters at Wright-Patterson AFB, Ohio.

An Iraqi Air Force MiG-25 Foxbat found buried under the sand at Al-Taqaddum Air Base, Iraq, 2003

In 1997, the United States bought 21 Moldovan aircraft for evaluation and analysis, under the Cooperative Threat Reduction accord, fourteen MiG-29Ss equipped with an active radar jammer in its spine and capable nuclear weapon armament. Part of the United States' motive to buy these aircraft prevented selling them to "rogue states," especially Iran. In late 1997, the MiG arrived at the National Air and Space Intelligence Center (NASIC) at Wright-Patterson Air Force Base near Dayton, Ohio.

In 2003, after the seizure of the Iraqi Air Force Al-Taqaddum Air Base, US troops found an advanced Russian MiG-25 Foxbat buried in the sand. US Air Force recovery exploitation teams dug the MiG out of a massive sand dune near the Al Taqqadum airfield.

The MiG was reportedly one of over two dozen Iraqi jets buried in the sand like a hidden treasure and recovered later. Reportedly, not all the jets found at captured Iraqi Air Force bases came from the Gulf War era.

The Russian-made MiG-25 Foxbat recovered was an advanced reconnaissance version never seen in the West and equipped with sophisticated electronic warfare devices. Air Force recovery exploitation teams used large earth-moving equipment to uncover the MiG, over 70 feet long and weighed 25 tons. The advanced electronic reconnaissance version found by the US Air Force is currently in service with the Russian Air Force.

US Designations and Fake Serial Numbers

Aircraft Receiving US Designations and Fake Serial Numbers to Identify Them in DOD Standard Flight Logs:
YF-110B MiG-21F-13 "Fishbed-C."
YF-110C A MiG-21 Chinese Chengdu J-7B (MiG-21F-13 A Fishbed variant)
YF-110D a MiG-21 "Fishbed" variant NATO

YF-112 Sukhoi Su-22 Fitter

YF-113A MiG-17F "Fresco-C" used in the HAVE DRILL program

YF-113B MiG-23BN "Flogger-F" NATO

YF-113C MiG-17F (a Chinese-built J-5) "Fresco-C" used in HAVE PRIVILEGE program

YF-113E MiG-23MS "Flogger-E" NATO

YF-114C MiG-17F "Fresco-C" (the one used in the HAVE FERRY program)

YF-114D MiG-17PF "Fresco-D NATO: (Serial: 75–008)

YF-116 MiG-25 Foxbat

YF-118 MiG-29 Falcrum

MiGs in Nevada exposed

In April 1984, Lt. Gen Robert M. Bond made two orientation flights in a Russian-built MiG-23 jet fighter. While making a high-speed run during his second flight, Bond lost control and crashed in Area 25 of the Nevada Test Site. He died while ejecting.

The 4477th flew its last MiG in 1990 and deactivated with the end of the Cold War. The 6513th which stopped flying MiG in the early eighties inactivated in 1992. In 2006, the Air Force declassified CONSTANT PEG program and held a series of press conferences about the former top-secret US MiG.

US MiGs flew more than 15,000 sorties, and nearly 7,000 aircrews flew in training against dissimilar aggressors in the Nevada desert between 1980 and the end of the program in 1988. These included the 'real' MiG, as well as the US-made A4 and T-38/F5 aircraft visually modified 'fake' MiG. The Groom Lake facility still exists (unofficially, of course).

Other countries taking the lead of the USAF and USN in aerial combat training

The Argentine Air Force trains its fighter pilots in the CEPAC (Course Combat Airman Standardization) course given at the IV Air Brigade in the town of "El Plumerillo" located in the province of Mendoza. The officers received from the Military Aviation School receive academic instruction and flight training aircraft in the AT-63 Pampa advanced and then integrate some of the operational squadrons of the Air Force.

The 410 Squadron of the Royal Canadian Air Force conducts an annual Fighter Weapons Instructor Course (FWIC) at CFB Cold Lake in Alberta. The course is three months in length and is specific to the CF-18 Hornet aircraft. There are eight students per course.

In Greece, the Hellenic Air Force built its school in 1975 called the Tactical Weapons School based in Andravida Air Base. In 1983, Hellenic Air Force established the KE.A.T. (Meaning Air Tactics Centre in Greek). Now the Tactical Weapons School is part of the KE.A.T. Both, based in Andravida, act as an independent Squadron of the Hellenic Air Force. Every year the best pilots from all the squadrons of the Hellenic Air Force train. The training encompasses modern air-to-air tactics, air-to-ground tactics, COMAO packages, and Electronic Warfare in KE.A.T. The pilots graduating are the best pilots in the Hellenic Air Force.

The Royal Netherlands Air Force has a Fighter Weapons Instructor Training (FWIT) with 323 Tactical Training, Evaluation & Standardization Squadron (TACTESS) at Leeuwarden Airbase as a multinational effort with Norway, Denmark, Belgium, and Portugal.

The Pakistan Air Force initially provided similar training via the fighter leader's school. The dedicated Combat Commanders School at PAF Base Mushaf replaced this in 1976. The Combat Commanders' School continues to fulfill its mission of training fighter pilots and air defense controllers. To keep pace with the ever-changing aerial threat and environment, CCS reviews its courses content continually. It adds new study and flying phases involving EW and BVR threat along with counterinsurgency tactics and others augmented to maintain pace with current Air warfare trends.

Turkish Air Force has a similar air-to-air and air-to-ground simulated warfare conditions training program based on highly developed ACMI (Air Combat Maneuvering Instrumentation) system, named Anatolian Eagle. Each year, several countries participate in operations, including Belgium, France, Germany, Israel, Italy, Jordan, Netherlands, United Arab Emirates, United Kingdom, and the United States of America.

Located in Konya 3rd. Main Jet Base, the main missions of the program included systematically testing and evaluating the fighters' combat readiness statuses and managing the tactical training progress.

The program built a background and knowledge base to make research on tactical aeronautics, to do research to allow fighter elements of the Turkish Air Force Command. The program reaches the military goals in the shortest time and with minimum resource and effort. It supports the definition of operational requirements and supply and R&D activities, allocating a training environment to fulfill the requirements of the Turkish Air Force Command to support the tests of existing/developed/future weapon/aircraft systems.

In the United Kingdom, the Royal Air Force and Royal Navy also had a similar course specific to each aircraft type, known as the QWI (Qualified Weapons Instructor, pronounced Que-Why) Course. It is five months in length.

The Russian Air Force also has its specialty course in modern air-to-air and air-to-ground combat. Based at Lipetsk Air Base, the Russians called it the 4th Center of Combat Application and Conversion of Frontline Aviation. It is under the command of General-Major Aleksander Kharchevsky. Lipetsk today has many of the old, current, and new Russian Air Force hardware, including the Sukhoi Su-34 and Yakovlev Yak-130.

Once the T-50 PAK-FA fifth-generation fighter comes into service, the first ten series production copies go to Lipetsk for the training of instructors. The new Sukhoi Su-35 will provide training for the new aircraft, as well as the T-50.

WORLD WIDE FISHBED AOB

(S-Gp-1)

NATIONAL AIR FORCES	MODEL C/E	D/F	UNK	TOTAL
Bulgaria	22	14		36
Czechoslovakia	42	76	16	134
E Germany	76	141		217
Hungary	60	30		90
Poland	43	68		111
Rumania	42	10		52
Yugoslavia	36	20		56
Russia		717		717
Communist China	33		13	46
North Korea	1	10	11	22
North Vietnam			6	6
Indonesia	17			17
Cuba	33	27		60
Syria	1			1
Iraq	11	13		24
India			64	64
United Arab Republic		36	72	108
Afganistan			32	32
	417	1162	214	1793

Soviet Air Forces in European Communist Countries

E Germany		284		284
Hungary		111		111
Poland		111		111
		506		506
TOTAL FISHBED - WORLD WIDE				2299

United States Navy TOPGUN

Immediately following the HAVE DOUGHNUT, HAVE DRILL, and HAVE FERRY exploitation projects at Area 51, the United States Navy recognized the need to train its pilots against the MiG planes they faced in SEA. The Navy recognized the need to teach its pilots how to fight the tactics used against them.

In May 1968, the Navy published the "Ault Report," which concluded the problem stemming from inadequate aircrew training in air combat maneuvering (ACM). The F-8 Crusader community is lobbying for an ACM training program ever since the Rolling Thunder began welcomed this. The Ault Report recommended the establishment of an "Advanced Fighter Weapons School" to revive and disseminate community fighter expertise throughout the fleet. CNO Moorer concurred.

On March 3, 1969, the Naval Air Station Miramar north of San Diego established the US Navy Weapons School. In 1996, the School merged into the United States Navy Strike Fighter Tactics Instructor program (SFTI program). Today, more popularly known as TOPGUN, the Naval Strike and Air Warfare Center at Naval Air Station Fallon, Nevada teaches fighter and strike tactics and techniques to selected Naval Aviators and Naval Flight Officers. Those trained returned to their operating units as surrogate instructors.

The Navy formed the school using many F-8 pilots as instructors, and placed under the control of the VF-121 "Pacemakers." The Pacemakers are an F-4 Phantom-equipped Replacement Air Group (RAG) unit, borrowing aircraft from its parent unit and other Miramar-based units.

Based on the Naval pilots first experiences with the MiG in Nevada, the objective developed, refined, and taught aerial dogfight tactics and techniques to selected fleet aircrews. It realistically replicated expected enemy aircraft widely used in air arms the world over. Then, the predominant enemy aircraft remained the Russian-built transonic MiG-17 'Fresco' and the supersonic MiG-21 'Fishbed' engaging the Naval pilots in SEA and recently engaged in Nevada.

The TOPGUN course chose aircrews from front-line units, who upon graduating, returned to their parent fleet units to relay what they had learned to their fellow squadron mates—becoming instructors themselves.

The HAVE DOUGHNUT and HAVE Drill tests provided the Navy data that led to the newly formed United States Navy Fighter Weapons School (TOPGUN) at NAS Miramar, California. What the Navy aviators learned changed the remainder of the Vietnam War, climbing the Navy kill to 8.33:1.

In contrast, the Air Force rate improved only slightly to 2.83:1. The reason for this difference was TOPGUN. The Navy (to include the Marine Corps) had revitalized its air combat training, while the Air Force remained stagnant. TOPGUN graduates made most of the Navy MiG kills. The Air Force, which had not implemented a similar training program, saw its kill ratio worsen.

The success of the US Navy fighter crews led to TOPGUN becoming a separate, fully funded command with its permanently assigned aviation, staffing, and infrastructural assets. Successful TOPGUN graduates scoring air-to-air kills over North Vietnam returned to instruct at TOPGUN. These included "Mugs" McKeown and Jack Ensch who flew against the MiGs in Nevada.

In 1996, NAS Miramar transferred to the Marine Corps. The Navy moved TOPGUN into the Naval Strike and Air Warfare Center (NSAWC) at NAS Fallon, Nevada. NAB Fallon continues to conduct an Adversary Training Course, flying with adversary aircrew much as it did during HAVE DOUGHNUT, HAVE DRILL, and HAVE FERRY.

An admiral commands the Naval Strike and Air Warfare Center at NAS Fallon is the Navy center of excellence for a naval strike and air warfare.

United States Air Force RED FLAG Exercises

Exploitation at Area 51 developed Air Force and Navy training programs

The Navy TOP GUN Fighter Weapons School and AF RED FLAG training exercises turned the kill ratio around. By the war's end; the US Navy/Marine Corps recorded 16 Jets lost to NVAF MiG while shooting down 61 NVAF MiG-17, MiG-19, and MiG-21s. The USAF records show losing 67 jets to the NVAF.

The NVAF in some writings might claim well over those figures. Be advised the NVAF MiG 17s & 19s might not all be Soviet-built aircraft, many of the MiG-17s Chicom J5 versions of the Soviet MiG-17F & the MiG-19s being Chicom J6s (Chinese Communist). The Soviets did NOT supply the NVAF with MiG-19s. The Red Chinese did NOT supply the NVAF with MiG-21s. However, they both supplied the NVAF with MiG-17s (and J5s).

While various studies and reports offered many suggestions, the fact is that Projects HAVE DOUGHNUT, HAVE DRILL, and HAVE FERRY revealed the reason for the lopsided aerial losses as the USAF and USN having lost the art of dogfighting as a skill. In the USAF, for instance, the solution in the 1960s to an increasing accident rate in the McDonnell Douglas F-4 Phantom II was simply to ban air combat maneuvering training (ACM). The accident rate fell. However, legions of Air Force Phantom drivers entered the skies of Vietnam with little experience in knowing what their aircraft could and could not do in a dogfight with a North Vietnamese MiG.

The United States Air Force did not react to what it learned during HAVE DOUGHNUT, HAVE DRILL, and HAVE FERRY as quickly as did the Navy. Consequently, the Air Force aircrews fighting the war in SEA continued to rate poorly. The Air Force showed the unacceptable performance of its fighter pilots and weapon systems in air combat maneuvering and air-to-air combat in comparison to previous wars.

Air combat over North Vietnam between 1965 and 1973 led to an overall 2.2:1 exchange ratio of enemy aircraft shot down to the number of own aircraft lost to enemy fighters. During Operation Linebacker, the ratio was less than 1:1. Among the several factors resulting in this disparity was a lack of realistic ACM training as learned when the Air Force encountered the MiG at Area 51 in Nevada.

It was not until 1975 that an Air Force analysis known as Project Red Baron II showed that a pilot's chances of survival in combat dramatically increased after he had completed ten combat missions. Thus, the United States Air Force created RED FLAG to offer USAF pilots and weapon systems officers the opportunity to fly ten realistically simulated combat missions in a safe training environment with measurable results. Many US aircrews had also fallen victim to SAMs during the Vietnam War, and RED FLAG exercises provided pilots and WSOs experience in this regime as well.

The concept of Colonel Richard "Moody" Suter became the driving force in RED FLAG's implementation, persuading the then-TAC commander, General Robert J. Dixon, to adopt the program. At Nellis, Suter was well known and well liked. The first RED FLAG exercise came off on Gen Dixon's schedule in November 1975. On 1 March 1976, the USAF chartered the 4440th Tactical Fighter Training Group (RED FLAG) with Col P. J. White as the first commander. Lt Col Marty Mahrt served as vice commander and Lt Col David Burner as Director of Operations. This small crew under Col White's leadership undertook the task of firmly establishing the program.

They selected the "aggressor squadrons," the opponents who flew against the pilots undergoing training, from the top fighter pilots in the US Air Force. They trained these pilots to fly per the tactical doctrines of the Soviet Union and other enemies of the period. This better simulated what then-TAC, as well as USAFE, PACAF and other NATO pilots and WSOs likely encountered in real combat against a Soviet, Warsaw Pact, or a Soviet-proxy adversary. The Air Force initially equipped the aggressors with readily available T-38 Talon aircraft to simulate the MiG-21, the T-38's similar size, and performance. F-5 Tiger II fighters, painted in color schemes commonly found on Soviet aircraft, added shortly after that became the mainstay until the introduction of the F-16.

In 1975, the United States Air Force conducted its first advanced aerial combat training exercise at Nellis Air Force Base, Nevada, known as RED FLAG. The US Air Force and its ally nations conduct several RED FLAG exercises during the year, each exercise lasting two weeks in duration. Since 1975, aircrews from the United States Air Force (USAF), US Navy (USN), United States Marine Corps (USMC), United States Army (USA) and numerous NATO or other allied nations' Air Forces have taken part in one or more exercises.

Under the aegis of the United States Air Force Warfare Center (USAFWC) at Nellis, the RED FLAG exercises continue today. Exercises are conducted in four to six cycles a year.

The 414th Combat Training Squadron (414 CTS) of the 57th Wing (57 WG) trains pilots and other flight crew members from the US, NATO, and other allied countries. They train for real air combat situations including the use of "enemy" hardware and live ammunition for bombing exercises within the adjacent Nevada Test and Training Range (NTTR).

At NTTR, the 414th Combat Training Squadron (RED FLAG) mission maximizes the combat readiness and survivability of participants. It provides a realistic training environment and a pre-flight and post-flight training within the Nellis Range Complex located northwest of Las Vegas and covering an area of 60 nautical miles.

During the Vietnam War, the US Navy and Air Force realized that a pilot surviving ten enemy encounters became experienced enough possibly to survive the war. Both the Navy and Air Force decided to give their pilots those ten sorties in the deserts of Nevada. To receive the needed training as realistic as possible, the adversary squadrons learned the aerial tactics of their enemies and used the enemies' tactics when engaging a Top Gun or Red Flag pilot in training.

Both the Navy and the Air Force uses stand-in aircraft to replicate expected enemy aircraft realistically and is widely used in air arms the world over. Their pilot's train with foreign pilots taught and who fight using the tactics of the enemy. Some claim the Red Eagles at Tonopah played the Russian national anthem, so the pilot thought Russian when engaging a pilot in training on the Nellis Gunnery Range.

In a typical RED FLAG exercise, Blue Forces (friendly) engage Red Forces (hostile) in realistic combat situations.

Blue Forces Are Made Up of Units from The Following;
- Air Combat Command (ACC),
- Air Mobility Command (AMC),
- Air Force Global Strike Command (AFGSC),
- Air Force Special Operations Command (AFSOC),
- The United States Air Forces Europe (USAFE),
- Pacific Air Forces (PACAF),
- Air National Guard (ANG),
- Air Force Reserve Command (AFRC),
- Air Force Space Command (AFSPC),
- Aviation units of the US Navy,
- US Marine Corps
- US Army,
- The Royal Air Force,
- Royal Canadian Air Force, and
- Royal Australian Air Force, as well as other allied Air Forces and fleet air arms.

A Blue Forces commander, who coordinates the units in an "employment plan" scheme of operation, leads them.

Red Forces (Adversary Forces) are composed of:
- The 57th Wing's 57th Adversary Tactics Group (57 ATG),
- Flying F-16s from the 64th Aggressor Squadron (64 AGRS)
- F-15s from the 65th Aggressor Squadron (65 AGRS) to provide realistic air threats through the emulation of opposition tactics.

Other US Air Force, US Navy, and US Marine Corps units flying in concert with the 507th Air Defense Aggressor Squadron's (507 ADAS) electronic ground defenses and communications, and radar jamming equipment also augment the Red Forces and provide GPS jamming.
- The 527th Space Aggressor Squadron (527 SAS),
- An Active Duty unit,
- The 26th Space Aggressor Squadron (26 SAS),
- An Air Force Reserve Command unit.

Additionally, the Red Force command and control organization simulate a realistic enemy integrated air defense system (IADS).

A key element of RED FLAG operations is the RED FLAG Measurement and Debriefing System (RFMDS). This is a computer hardware and software network that provides real-time monitoring, post-mission reconstruction of maneuvers and tactics providing participant pairings and integration of range targets and simulated threats. Blue Force commanders objectively assess mission effectiveness and validate lessons learned from data provided by the RFMDS.

A typical flag exercise year includes ten Green Flags, a close air support (CAS) exercise with the US Army, one Canadian Maple Flag (operated by the Royal Canadian Air Force), and four RED FLAGs. Each RED FLAG exercise normally involves a variety of fighter interdiction, attack/strike, air superiority, enemy air defense suppression, airlift, air refueling, and reconnaissance missions. In a 12-month period, more than 500 aircraft fly more than 20,000 sorties, while training more than 5,000 aircrews and 14,000 support and maintenance personnel.

Before a "flag" begins, the RED FLAG staff conducts a planning conference where unit representatives and planning staff members develop the size and scope of their participation. All aspects of the exercise, including billeting of personnel, transportation to Nellis AFB, range coordination, ordnance/munitions scheduling, and development of training scenarios, are designed realistic, fully exercising each participating unit's capabilities and objectives.

Today, the 414th Combat Training Squadron (414 CTS) is the unit currently tasked with running RED FLAG exercises. The 64th Aggressor Squadron (64 AGRS) and the 65th Aggressor Squadron (65 AGRS) also based at Nellis AFB use F-16, and F-15 aircraft Flanker painted in the various camouflage schemes of potential adversaries. They emulate, respectively, the MiG-29 Fulcrum and Su-30.

The US Air Force's RED FLAG approach differs from that initially employed during and after the Vietnam War by the US Navy to improve fighter aircrew performance. Rather than a large, multi-squadron exercise, the Navy established the US Navy Fighter Weapons School (more widely known as TOPGUN) in 1969 at the former NAS Miramar, California. TOPGUN "trained the trainers," with Navy and Marine Corps squadrons in the Fleet selecting their best fighter, strike fighter, and Marine fighter/attack aircrews for training when back at their home stations between overseas deployments. TOPGUN graduates then returned to their Fleet squadrons before their next overseas deployment to share lessons learned with their fellow Naval Aviators and Naval Flight Officers.

Master Jet Bases later established Navy and Marine Corps adversary squadrons. These bases include NAS Miramar (now MCAS Miramar), NAS Oceana, NAS Lemoore, MCAS Yuma, MCAS Beaufort, the former NAS Cecil Field, the former MCAS El Toro. The bases also include fleet air training bases. These are bases such as NAS Fallon, NAS Key West, and the former NS Roosevelt Roads. Then include forward deployed bases such as the former NAS Cubi Point, for Fleet squadrons to conduct dissimilar air combat training (DACT) as part of unit level training (ULT).

These adversary squadrons initially flew the A-4 Skyhawk, with the Navy and Marine Corps. They later added the T-38 Talon, and F-5E/F to its adversary lineup and briefly included the F-21 Kfir. Other naval adversary aircraft have included specially built F-16Ns, the F-14, and the F/A-18. Today, Carrier Air Wing level training, analogous to the USAF RED FLAG program, is conducted at NAS Fallon. There the Naval Strike and Air Warfare Center (NSAWC), of which TOPGUN in now part, operates dissimilar adversary aircraft (F-16 and F/A-18). A collocated Naval Air Reserve squadron, Fighter Composite Squadron 13 (VFC-13), flies the F-5E and F-5F.

The United States Marine Corps (USMC) also conducts Weapons, and Tactics Instructor (WTI) exercises at Marine Corps Air Station Yuma twice a year as part of the WTI course conducted by Marine Air Weapons and Tactics Squadron ONE (MAWTS-1). The Marine Corps uses locally based Marine Fighter Training Squadron 401 (VMFT-401), a Marine Air Reserve squadron and the only USMC adversary squadron.

Originally equipped with the F-21 Kfir, VMFT-401 now operates the F-5E and F-5F.

In 2009, the 416th Flight Test Squadron (416 FLTS) from Edwards AFB, California, also participated in RED FLAG, the first time for an Air Force Material Command (AFMC) unit to take part in the program.

In the Southeast Asian conflict, however, that exchange ratio fell to less than 1-to-1 during a period in the spring of 1972. Today's RED FLAG over a single year involves as many as 250 different units and 750 aircraft of many different types. About 11,000 aircrews and squadron personnel amass more than 12,000 sorties and 21,000 flight hours during the year. One major milestone in that history, without question, was the stunning performance of American airmen in the Gulf War of 1991. It was the first war to highlight the results of RED FLAG. It produced a unique tribute when an Air Force pilot returning from a combat mission over Iraq remarked, "It was almost as intense as RED FLAG."

While many programs at Area 51 remain classified, as an epilogue to what the CIA initiated at Area 51 is slowly coming to light through information released to the public through formal announcements, published technical papers, and official personnel biographies that reveal details of previously "black" projects, declassified activities and programs such as technology demonstrators used to pioneer revolutionary advances in low-observables, aircraft design, and rapid prototyping. Hundreds of sorties were flown, many of these aircraft being foreign types but some possibly entirely new aircraft. There have been at least seven and possibly as many as 11 classified manned aircraft flown at Groom Lake since the mid-1980s that have yet to be unveiled. This doesn't include the modified aircraft, foreign aircraft, or ordinary platforms (C-130, F-16, etc.) carrying experimental avionics, or unmanned classified aircraft. One can only wonder what is occurring in Area 51 now and will occur in the future.

Participating Countries

Only countries considered friendly towards the United States take part in RED FLAG exercises. So far, the countries that have been involved in these exercises are:

- Australia
- Belgium
- Brazil (July 1998, August 2008, and February 2013)
- Canada
- Chile (July 1998)
- Colombia (July 2012)
- Denmark
- Egypt
- France
- Germany
- Greece (October 2008)
- India
- Israel
- Italy
- Japan
- New Zealand

- Netherlands
- Norway
- Pakistan (2010)
- Poland (June 2012)
- Portugal (March 2000)
- Sweden
- Singapore
- Saudi Arabia
- South Korea
- Spain
- Thailand
- Turkey
- The United Arab Emirates
- United Kingdom
- Venezuela

****

The MiG Bandits of Nevada

Red Hats (HAVE DOUGHNUT, HAVE Drill/Ferry, Have Idea) 1968–1993
Robert "Bob" Acosta (1986–1989) – AFSC, Have Idea
Skip Anderson (1973–1976) – AFSC, Have Idea
Robert G. Ashcraft (1968–1969) – TAC, HAVE DOUGHNUT, HAVE Drill/Ferry
John J. Barnoski (1987–1991) – AFSC, Have Idea, 6513TS/CC
Jon S. Beesley (1980–1981) – AFSC, Have Idea
Robert Belt (1989) – TAC, Have Idea
Matt Black (1992–1993) – ACC, Have Idea
James E. "J. B. " Brown III (1989–1993) – AFSC/AFMC, Have Idea
David R. Bryant (1983–1986) – USN, Have Idea
Joe Lee Burns (1972–1974) – TAC, Have Idea
Kevin P. Burns (1989–1993) – AFSC/AFMC, Have Idea, 6513TS/CC, 413FLTS/CC
David Carr (1993) – ACC, Have Idea
John H. Casper (1975–1981) – AFSC, Have Idea, 6513TS/CC
Thomas J. Cassidy, Jr. (1968) – USN, HAVE DOUGHNUT
Chris Ceplecha (1991–1993) – TAC/ACC, Have Idea
Philip J. Conley, Jr. (1978–1982) – AFSC, Have Idea, AFFTC/CC
Fred J. Cuthill (1968–1969) – AFSC, HAVE DOUGHNUT, HAVE Drill/Ferry
Daniel N. Dixon (1992–1993) – USN, Have Idea
Norman K. "Ken" Dyson (1976–1977) – AFSC, Have Idea
John R. "Russ" Easter (1981) – AFSC, Have Idea
David L. Ferguson (1973–1979) – AFSC, Have Idea, 6513TS/CC
Peter B. Field USMC, Have Idea
Mike Ford TAC, Have Idea
Robert S. Frank (1985–1988) – AFSC, Have Idea, 6513TS/CC
Ralph H. Graham (1981–1985) – AFSC, Have Idea, DET3/CC
Steven A. Green (1985–1986) – AFSC, Have Idea
Thomas C. Horne (1993–1995) – AFMC, 413 FLTS/CC
Jerry "Devil" Houston (1969) – USN, HAVE Drill/Ferry
Paul P. Jacobs, Jr. AFSC, Have Idea
James A. Jimenez (1991–1993) – AFMC
Maurice B. "Duke" Johnson (1969) – TAC, HAVE Drill/Ferry
Gayland E. Jones AFSC, Have Idea
Joe B. Jordan (1968) – AFSC, HAVE DOUGHNUT
George Kailiwai III AFSC, Have Idea
D. Keator (1980) TAC, Have Idea
John V. Kelley AFSC, Have Idea
William J. "Pete" Knight – AFSC, Have Idea
Fred D. Knox, Jr. (1981–1983) – USN, Have Idea
Joseph A. "Broadway Joe" Lanni (1992–1995) – AFMC, Have Idea
Gerald D. Larson (1968) – TAC (1137th SAS) HAVE DOUGHNUT
Rodney K. H. Liu – AFSC, Have Idea
Michael V. Love – AFSC, Have Idea
James D. "Tony" Mahoney (1990–19??) – TAC/ACC, Have Idea
Marvin L. "Roy" Martin (1979–1981) – AFSC, Have Idea
David G. Mazur –AAFSC, Have Idea
Larry D. McClain (1977–1981) – AFSC, Have Idea
Ronald E. "Mugs" McKeown (1969) – USN, HAVE Drill/Ferry
Donald R. McMonagle (1986–1987) – AFSC, Have Idea

Edward T. "Tom" Meschko – AFSC, Have Idea

Robert A. Moseley (1981) – AFSC, Have Idea

George K. Muellner (1979–1982) – AFSC, Have Idea

Thomas A. Morganfeld (1975–1976) – USN, Have Idea

Steven R. Nagel (1977–1979) – AFSC, Have Idea

William E. Nelson AFSC, Have Idea

Dennis L. Nuttbrock AFSC, Have Idea

Craig T. Otto AFSC, Have Idea

Philip C. Pirozzi (1991–1992) – USN, Have Idea

Michael C. "Bat" Press (1972–1977) – TAC, Have Idea

Joe M. Roberts AFSC, Have Idea

Dennis F. Sager (1991–1993) – AFSC/AFMC, Have Idea

Paul W. Savage (1986) – AFSC, Have Idea

Wendell H. Shawler (1969–1970) – AFSC, HAVE Drill/Ferry, Have Privilege

David L. "DL" Smith (1972–1975) – TAC, Have Idea

Thomas P. Stafford (1975–1978) – AFSC, Have Idea, AFFTC/CC

Norman L. Suits (1971–1973) – AFSC, Have Idea

Thomas S. Swalm (1969) – TAC, HAVE Drill/Ferry

Dane C. Swanson (1985–1989) – USN, Have Idea

Paul D. Tackabury (1980–1981) – AFSC, Have Idea

John A. "Ashby" Taylor (1981–1985) – AFSC, Have Idea, 6513TS/CC

Foster S. "Tooter" Teague (1969) – USN, HAVE Drill/Ferry

James W. Tilley II (1980–1984) – AFSC, 6513TS/CC

William T. "Ted" Twinting (1968) – AFSC, HAVE DOUGHNUT

Teddy Varwig (1991–1992) – TAC, Have Idea

Ken Wallace (1988) – USN, Have Idea

Charles P. "Pete" Winters (1971–1983) – AFSC, Have Idea, DET3/CC

Otto J. Waniczek (1979–1982) – AFSC, Have Idea, FTE

Mike Welch (1969) – USN HAVE Drill/Ferry

James H. Wisneski (1981–1986) – AFSC, Have Idea

Red Eagles (CONSTANT PEG, Have Idea) 1977–1988

David F. "Blazo" Bland (1983–1985) – TAC, CONSTANT PEG (Bandit 32)

Thomas V. "Boomer" Boma (1986–19??) – TAC, CONSTANT PEG (Bandit 59)

Melvin Hugh "Bandit" Brown (1979) – USN, CONSTANT PEG (Bandit 12)

Steven R. "Brownie" Brown (1983–1986) – TAC, CONSTANT PEG (Bandit 33)

Guy A. Brubaker (1984–1986) – USN, CONSTANT PEG (Bandit 48)

Leonard J. Bucko (1981–1983) – USMC, CONSTANT PEG (Bandit 22)

Herbert J. "Hawk" Carlisle (1986–1988) – TAC, CONSTANT PEG (Bandit 54)

Ricardo M. "Rick" Cazessus (1986–19??) – TAC, CONSTANT PEG (Bandit 63)

Charles T. "Chuck" Corder (1979–1983) – TAC, CONSTANT PEG (Bandit 15)

Evan M. Chanik, Jr. (1985–1988) – USN, CONSTANT PEG (Bandit 52)

Robert E. Craig (1984–1987) – TAC, CONSTANT PEG (Bandit 43)

Robert E. "Sundance" Davis (1987–198?) – USN, CONSTANT PEG (Bandit 64)

Thomas E. "Gabby" Drake (1984–1987) – TAC, CONSTANT PEG (Bandit 42)

James B. "Bear" Evans (1987–198?) – TAC, CONSTANT PEG (Bandit 67)

Glenn F. "Pappy" Frick (1972–1978) – TAC, Have Idea, CONSTANT PEG, 4477TEF/CC (Bandit 4)

Nickie J. Fuerst (1987–198?) – TAC, CONSTANT PEG (Bandit 65)

Francis K. "Paco" Geisler (1983–1986) – TAC, CONSTANT PEG (Bandit 35)

George S. "G2" Gennin (1982–1984) – TAC, CONSTANT PEG, 4477TES/CC (Bandit 31)

Thomas A. Gibbs (1980–1982) – TAC, CONSTANT PEG, 4477TES/CC (Bandit 21)

James W. "Wiley" Green (1982–1984) – TAC, CONSTANT PEG (Bandit 26)

Charles J. "Heater" Heatley III (1977–1981) – USN, CONSTANT PEG (Bandit 8)

Earl J. Henderson (1979–1980) – USN, Have Idea, CONSTANT PEG 4477TEF/CC, 4477TES/CC (Bandit 14)

Gerrald D. "Huffer" Huff (1977–1980) – TAC, CONSTANT PEG (Bandit 6)

Ronald W. Iverson (1975–1979) – TAC, Have Idea (Bandit 2)

Timothy R. "Stretch" Kinney (1984–1987) – TAC, CONSTANT PEG (Bandit 45)

Dudley A. Larsen (1986–19??) – TAC, CONSTANT PEG (Bandit 60)

Selvyn S. "Sel" Laughter (1980–1982) – USN, CONSTANT PEG (Bandit 18)

Martin S. Macy (1985–1988) – USMC, CONSTANT PEG (Bandit 49)

James D. "Tony" Mahoney (1987–1990) – USN, CONSTANT PEG (Bandit 62)

John T. "Jack" Manclark (1985–1987) – TAC, CONSTANT PEG, 4477TES/CC (Bandit 51)

John C. "Flash" Mann (1986–1988) – TAC, CONSTANT PEG (Bandit 56)

James D. "Thug" Matheny (1982–1985) – TAC, CONSTANT PEG (Bandit 27)

Robert "Kobe" Mayo (1972–1976) – TAC, Have Idea (Bandit 1)

David J. "Marshall" McCloud (1979–1981) – TAC, CONSTANT PEG (Bandit 6)

Daniel R. "Bad Bob" McCort (1984–1986) – USN, CONSTANT PEG (Bandit 44)

Douglas M. Melson (1987–1988) – TAC, CONSTANT PEG (Bandit 68)

Thomas A. Morganfeld (1977–1980) – USN, Have Idea, CONSTANT PEG (Bandit 7)

Brian E. McCoy (1986–19??) – TAC, CONSTANT PEG (Bandit 53)

Alvin D. "Devil" Muller (1974–1978) – TAC, Have Idea (Bandit 3)

Clem B. "Buffalo" Myers (1980–19??) – TAC, CONSTANT PEG (Bandit 19)

John B. "Black" Nathman (1982–1984) – USN, CONSTANT PEG (Bandit 29)

Joseph L. "Jose" Oberle (1978–1982) – TAC, CONSTANT PEG (Bandit 5)

Stanley R. "Swish" O'Connor (1988–1990) – USN, CONSTANT PEG (Bandit1969)

Gaillard R. "Evil" Peck (1978 – 1979) – TAC, CONSTANT PEG, 4477TEF/CC (Bandit 9)

Edward D. "Hog" Phelan (1984–1988) – TAC, CONSTANT PEG (Bandit 47)

Mark F. "Toast" Postai (1981–1982) – TAC, CONSTANT PEG (Bandit 25)

Michael C. "Bat" Press (1972–1977, 1980–1981) – TAC, Have Idea, CONSTANT PEG (Bandit 20)

Orville Prins (1982–1985) – TAC, CONSTANT PEG (Bandit 30)

James A. "Rookie" Robb (1983–1985) – USN, CONSTANT PEG (Bandit 38)

Shelley S. "Scotty" Rogers (1986–19??) – TAC, CONSTANT PEG (Bandit 55)

Michael C. Roy (1983–1985) – TAC, CONSTANT PEG (Bandit 36)

John B. Saxman (1983–1986) – TAC, CONSTANT PEG (Bandit 34)

Michael R. "Scotty" Scott (1979–1983, 1987–1990) – TAC, CONSTANT PEG, 4477TES/CC (Bandit 14)

Keith E. Shean (1980–1982) – TAC, CONSTANT PEG (Bandit 17)

Robert "Catfish" Sheffield (1979–1982) – TAC, CONSTANT PEG (Bandit 16)

Larry T. "Shy" Shervanick (1982–1985) – TAC, CONSTANT PEG (Bandit 28)

Cary A. "Dollar" Silvers (1986) – USN, CONSTANT PEG (Bandit 61)

Michael G. Simmons (1986–19??) – TAC, CONSTANT PEG (Bandit 57)

John "Grunt" Skidmore (1984–1985) – TAC, CONSTANT PEG (Bandit 41)

Paul R. "Stook" Stucky (1983–1986) – TAC, CONSTANT PEG (Bandit 40)

Charles E. "Smokey" Sundell (1986–19??) – TAC, CONSTANT PEG (Bandit 58)

Russell M. "Bud" Taylor (1981–1983) – USN, CONSTANT PEG (Bandit 23)

Sam C. Therrien (1987–198?) – TAC, CONSTANT PEG (Bandit 66)

Frederick H. "T-Bear" Thompson (1985–198?) – TAC, CONSTANT PEG (Bandit 50)

George C. "Cajun" Tullos (1983–1985) – USMC, CONSTANT PEG (Bandit 37)

James M. "Monroe" Watley (1981–1984) – TAC, CONSTANT PEG (Bandit 24)

Philip W. "Hound Dog" White (1984–1986) – TAC, CONSTANT PEG, 4477TES/CC (Bandit 46)

Karl F. "Harpo" Whittenberg (1979–19??) – TAC, CONSTANT PEG (Bandit 11)

US planes lost in Vietnam

The United States Air Force

All told, the US Air Force flew 5.25 million sorties over South Vietnam, North Vietnam, northern and southern Laos, and Cambodia, losing 2,251 aircraft: 1,737 to hostile action and 514 in accidents. 110 of the losses were helicopters and the rest fixed-wing. A ratio of roughly 0.4 losses per 1,000 sorties compared favorably with a 2.0 rate in Korea and the 9.7 figure during World War II.

Sources for USAF Figures:

USAF Operations Report, 30 November 1973

Campbell, John M., and Hill, Michael. Roll Call: Thud. Atglen, PA: Schiffer Publishing Ltd., 1996. ISBN 0–7643–0062–8.

Hobson, Chris. Vietnam Air Losses, USAF, USN, USMC, Fixed-Wing Aircraft Losses in Southeast Asia 1961–1973. North Branch, Minnesota: Specialty Press, 2001. ISBN 1–85780–115–6.

USAF fixed-wing

Downed USAF Douglas A-1E, pilot later awarded the Medal of Honor

A-1 Skyraider— —191 total, 150 in combat

First loss A-1E 52–132465 (1st Air Commando Squadron, 34th TG) shot down during the night of 28–29 August 1964 near Bien Hoa, SVN

Final loss A-1H 52–139738 (1st Special Operations Squadron, 56th Special Operations Wing) shot down 28 September 1972 (pilot rescued by an Air America helicopter).

A-7D Corsair II— —6 total, 4 combat

–First loss 71–0310 (355th Tactical Fighter Squadron, 354th TFW) on 2 December 1972 shot down on a CSAR mission in Laos (Capt Anthony Shine KIA).

–71–0312 (353d TFS) mid-air collision with an FAC O-1 Bird Dog in Laos on 24 December 1972, (Capt Charles Riess POW)

-71–0316 (355th TFS) operational loss (non-combat) crash in Thailand on 11 January 1973 (Pilot Rescued)

-70–0949 (354th TFW) shot down Laos on 17 February 1973 (Maj J. J. Gallagher Rescued)

-71–0305 (3rd TFS, 388th TFW) shot down in Cambodia on 4 May 1973 (1Lt T. L. Dickens Rescued)

Final loss 70–0945 (354th TFW) shot down in Cambodia on 25 May 1973 (Capt Jeremiah Costello KIA)

A-26 Invader— —22 total

–First loss B-26B 44–35530 (Detachment 2A, 1st ACG) shot down in IV CTZ on the night of 4–5 November 1962 killing the 3 crew.

–Final loss A-26A 64–17646 (609th SOS, 56th SOW) lost over Laos on the night of 7–8 July 1969 killing both crew.

A-37 Dragonfly— —22 total

–First loss 1967; final loss 1972

Wing of downed USAF warplane

AC-47 Spooky— —19 total, 12 in combat

–First loss 1965, final loss 1969

AC-119 Shadow/Stinger— —6 total, 2 in combat

–First loss AC-119G 52–5907 (Det.1, 17th SOS, 14th SOW) which crashed on take-off from Tan Son Nhut, SVN on 11 October 1969 killing 6 of the 10 crew.

–Final loss 1971

AC-130 Spectre— —6 total, all combat.

-First loss AC-130A 54–1629 (16th SOS, 8th TFW) hit by 37 mm AAA over Laos and crash-landed at Ubon RTAFB, 2 crew died (one died of injuries before reaching Ubon) however, 11 others survived.

-Final loss 1972

B-52 Stratofortress— —31 total, 17 in combat

-First losses were operational (non-combat) mid-air collision 2 B-52F 57–0047 and 57–0179 (441st Bomb Squadron, 320th Bomb Wing), 18 June 1965, the South China Sea during air refueling orbit, 8 of 12 crew killed

-Final loss B-52D 55–0056 (307th Bomb Wing Provisional) to SAM 4 January 1973, crew rescued from Gulf of Tonkin

B-57 Canberra— —56 total, 38 in combat

-First loss 1964, final loss 1970

C-5A Galaxy— —1 total, 0 in combat. Crashed while attempting emergency landing at Tan Son Nhut Air Base 4 April 1975, as part of Operation Babylift Five of the eight US Military women killed during the Vietnam War, was aboard this aircraft

C-7 Caribou— —19 total, 9 in combat

-First lost C-7B 62–4161 (459th Tactical Airlift Squadron, 483d Tactical Airlift Wing) hit by a US 155 mm shell on 3 August 1967 in SVN killing the 3 crew. Note: two fatal crashes occurred during Operation Red Leaf transition training of USAF crews in Army CV-2's, on 4 and 28 October 1966.

-Final loss was C-7B 62–12584 (483d TAW) which crashed in SVN on 13 January 1971, all 4 crew survived.

C-47 Skytrain— —21 total

-A C-47 was very first USAF aircraft lost in the SEA conflict, C-47B 44–76330 (315th Air Division) on TDY at Vientiane, Laos shot down by the Pathet Lao on 23 March 1961 killing 7 of the 8 crew. The sole survivor, US Army Maj Lawrence Bailey was captured and held until August 1962.

-Final loss EC-47Q 43–48636 (361st Tactical Electronic Warfare Squadron, 56th SOW) shot down in Laos on the night of 04/5 4–5 February 1973 killing all eight crew.

C-123 Provider— —53 total, 21 in combat

-First loss was C-123B 56–4370 attached to the 464th TAW which came down on an Operation Ranch Hand (defoliation) training flight between Bien Hoa and Vung Tau, SVN on 2 February 1962

-Final loss 1971

C-130 Hercules— —55 total, 34 in combat

The first loss was C-130A 57–0475 (817th Troop Carrier Squadron, 6315th Operations Group). On 24 April 1965, a Blind Bat flareship crashed into high ground near Korat Royal Thai Air Force Base, Thailand while attempting to land in bad weather with a heavy load. Two engine failures and low fuel killed all six crew, the 14th recorded loss of a C-130 to all causes.

-Final loss C-130E 72–1297 (314th TAW) destroyed by rocket fire at Tan Son Nhut Air Base on 28 April 1975.

C-141 Starlifter— —2 total, 0 combat

-C-141A 65–9407 (62d Military Airlift Wing) destroyed in a night runway collision with a USMC A-6 at Danang, SVN on 23 March 1967 killing 5 of the 6 crew.

-C-141A 66–0127 (4th Military Airlift Squadron, 62d MAW) crashed soon after take-off from Cam Ranh Bay, SVN on 13 April 1967 killing 6 of the 8-man crew.

E/RB-66 Destroyer—14 total

-First loss was RB-66B 53–0452 (Det 1, 41st Tactical Reconnaissance Squadron, 6250th Combat Support Group) which crashed 22–23 October 1965 west of Pleiku, SVN killing the crew.

-Final loss EB-66B 42nd TEWS, 388th TFS lost to engine failure on 23 December 1972 during Operation Linebacker II. 3 crew were KIA.

EC-121 BatCat— —2 total, 0 combat

-EC-121R 67–24193 (554th Reconnaissance Squadron, 553d RW) crashed 25 April 1969 on take-off in a thunderstorm from Korat RTAFB, killing all 18 crew.

-EC-121R 67–21495 (554th RS) crashed on approach to Korat RTAFB on 6 September 1969, 4 of the 16 men killed.

F-4 Phantom II— —445 total, 382 in combat

-The first loss was operational (non-combat), F-4C 64–0674 (45TH TFS, 15th TFW) which ran out of fuel after the strike in SVN on 9 June 1965; first combat loss F-4C 64–0685 (45th TFS, 15th TFW) shot down Ta Chan, NW NVN on 20 June 1965. 9 of the losses were parked aircraft struck by rockets.

-Final loss, F-4D 66–8747 (432d TRW) on 29 June 1973.

F-5 Freedom Fighter— —9 total

-First loss 1965, final loss 1967

F-100 Super Sabre— —243 total, 198 in combat

-First loss 1964, final loss 1971

F-102 Delta Dagger— —14 total, 7 combat

-First loss 1964, final loss 1967. 4 of the combat losses were parked aircraft

F-104 Starfighter— —14 total, 9 combat

-First loss 1965, final loss 1969

F-105D Thunderchief— —335 total, 283 in combat

-First loss 62–4371 (36th TFS, 6441st TFW) written off from battle damage over Laos 14 August 1964, at Korat, Thailand

-Final loss 61–0153 (44th TFS, 355th TFW) shot down Laos 23 September 1970, pilot Capt J. W. Newhouse rescued

F-105F/G Thunderchief: "Wild Weasel," "Ryan's Raiders," "Combat Martin"— —47 total, 37 combat

-First loss EF-105F 63–8286 (13th TFS, 388th TFW) shot down by AAA RP-6 July 1966, Maj Roosevelt Hestle, and Capt Charles Morgan KIA

-Last loss F-105G 63–8359 (Det.1 561st TFS, 388th TFW) shot down by SAM 16 November 1972, RP-3, crew rescued

F-111A "Aardvark"— —11 total, 6 in combat

-First loss mission-related TFR failure, 66–0022 (428th TFS 474th TFW, Project Combat Lancer), 28 March 1968, Maj H.E. Mccann and Capt D.L. Graham MIA

-Final loss 67–0111 (474th TFW) mid-air collision over Cambodia, 16 June 1973, both crews rescued

HU-16 Albatross— —4 total, 2 combat

-51–5287 to unknown cause 19 June 1965

-51–0058 to unknown cause 3 July 1965

-51–0071 (33d ARRS) shot down by AAA 14 March 1966, two crew killed

-51–7145 (37th ARRS) disappeared 18 October 1966, 7 crew KIA-BNR

KB-50 Superfortress tanker— —1 total, 0 combat

-Only loss KB-50J 48–0065 (421st Air Refueling Squadron Detachment) at Takhli RTAFB, which crashed in Thailand on 14 October 1964, all 6 crew, survived.

KC-135 Stratotanker— —3 total, 0 combat

-Two crashes in 1968, one 1969, all operational (non-combat)

O-1 Bird Dog— —172 total, 122 in combat

-First loss 1963, final loss 1972

O-2 Skymaster— —104 total, 82 in combat

-First loss 1967, final loss 1972

OV-10 Bronco— —63 total, 47 in combat

First loss 1968, final loss 1973

QU-22 Pave Eagle— —8 lost, 7 in combat

-First loss YQU-22A 68–10531 (554th RS, 553d RW) crashed due to engine failure on 11 June 1969

-Final loss QU-22B 70–1546 (554th RS) on 25 August 1972, pilot killed.

RF-4C Phantom II— —83 total, 76 in combat

-First loss 1966, final loss 1972

RF-101 Voodoo— —39 total, 33 in combat

-First loss 1964, final loss 1968

SR-71A Blackbird— —2 total, 0 combat

-64–17969 (Det OL-8, 9th Strategic Reconnaissance Wing) suffered engine failure over Thailand on 10 May 1970, both crews ejected safely

-64–17978 (Det OL-KA, 9th SRW) crashed on landing at Kadena, Okinawa on 20 July 1972, both crews survived

T-28 Trojan— —23 total

-First loss 1962, final loss 1968

U-2C "Dragon Lady"— —1 total, 0 combat

-Only loss 56–6690 (349th Strategic Reconnaissance Squadron 100th SRW) which crashed on 8 October 1966 near Bien Hoa, SVN, Maj Leo J Stewart ejected and rescued.

U-3B Blue Canoe— —1 total, 1 combat

-Only loss 60–6058, destroyed on the ground during a VC attack on Tan Son Nhut, SVN on 14 June 1968.

U-6A Beaver— —1 total, 0 combat

Only loss 51–15565 (432d Tactical Reconnaissance Wing) which crashed in Thailand 28 December 1966, both crews survived.

U-10D Courier— —1 total, 1 combat

-63–13102 (5th SOS, 14th SOW) shot down 14 August 1969 near Bien Hoa killing 1/Lt Roger Brown.

CH/HH-3 Jolly Green Giant— —34 total, 25 in combat

-First loss CH-3E 63–9685 (38th ARRS) to AAA North Vietnam on 6 November 1965, three crew POW, one rescued

-Last loss HH-3E 65–12785 (37th ARRS) 21 November 1970, combat-assaulted inside Son Tay POW camp (Operation Ivory Coast) and deliberately destroyed by US Special Forces

HH-43B Pedro— —13 lost, 8 in combat

-First loss 63–9713 (38th ARRS) damaged by fire 2 June 1965, crew rescued and aircraft destroyed to prevent its capture

-Final loss 60–0282 (38th ARRS) crashed Cam Ranh Bay 7 August 1969, crew rescued

CH/HH-53 Super Jolly— —27 total, 17 in combat

-First loss HH-53C 66–14430 (40th ARRS) in Laos, damaged by gunfire 18 January 1969 crew rescued and aircraft destroyed by bombing to prevent capture

-Last losses four CH-53's (68–10925, –10926, –10927, 70–1627 all from 21st SOS, 56th SOW) to AAA on 15 May 1975, Koh Tang, Kampuchea, (Mayaguez incident final aircraft losses of Vietnam War)

UH-1 Iroquois— —36 total

US Navy

Twenty-one aircraft carriers conducted 86 war cruises and operated 9,178 total days on the line in the Gulf of Tonkin. 530 aircraft were lost in combat and 329 more to operational causes. Resulting in the deaths of 377 naval aviators, with 64 airmen reported missing and 179 taken prisoner-of-war.

Sources for USN carrier-based figures:

Francillon, René. Tonkin Gulf Yacht Club: US Carrier Operations off Vietnam, Naval Institute Press (1988) ISBN 0–87021–696–1

USN fixed-wing carrier-based

A-1 Skyraider—65 total, 48 in combat

-First loss A-1H 139760 (VA-145, USS Constellation), to AAA 5 August 1964, Lt.j.g. R. C. Sather KIA (Body recovered in 1985)

-Final loss A-1H 134499 (VA-25, USS Coral Sea), to MiG 14 February 1968, Lt.j.g. J. P. Dunn MIA

A-3 Skywarrior—7 total, 2 in combat

-First loss A-3B 142250 (VAH-4, USS Hancock), operational loss (non-combat) 22 December 1964, 3 rescued, 1 killed

-Final loss A-3B 144627 (VAH-4, USS Kitty Hawk), AAA 8 March 1967, 3 crew KIA

A-4 Skyhawk—282 total, 195 in combat

-First loss A-4C 149578 (VA-144, USS Constellation), AAA 5 August 1964, Lt.j.g. Everett Alvarez POW (second longest held prisoner)

-Final loss A-4F 155021 (VA-212, USS Hancock), AAA 6 September 1972, pilot rescued

A-6 Intruder—62 total, 51 in combat

-First loss A-6A 151584 (VA-75, USS Independence), own bomb detonation Laos 14 July 1965, crew rescued

-Final loss A-6A 157007 (VA-35, USS America), AAA South Vietnam 24 January 1973, crew rescued

A-7 Corsair—100 total, 55 in combat

-First loss A-7A 153239 (VA-147, USS Ranger), SAM North Vietnam 22 December 1967, LCDR J.M. Hickerson POW

-Final loss A-7E 156837 (VA-147, USS Constellation), operational loss (non-combat) 29 January 1973, pilot missing

C-1 Trader—4 total, 0 in combat

-C-1A 146047 (VR-21, USS Independence), non-combat 30 August 1965, 7 passengers and crew rescued

-C-1A 136784 (VR-21, USS Independence), operational loss (non-combat) 12 September 1965, 9 passengers and crew rescued, 1 killed

-C-1A 146016 (Composite Squadron Five VC-5), operational loss (non-combat) 8 August 1967, 3 passengers and 2 crew rescued

-C-1A 146054 (Carrier Air Wing 11, Kitty Hawk), operational loss (non-combat) 16 January 1968, 7 passengers and crew rescued, 3 killed

C-2 Greyhound—1 total, 0 in combat

-Sole loss C-2A 155120 (VRC-50, USS Ranger), Gulf of Tonkin crash 15 December 1970, 9 passengers and crew killed

E-1 Tracer—3 total, 0 in combat

-First loss E-1B 148918 (VAW-12, USS Independence), operational loss (non-combat) 22 September 1965, crew rescued

-Final loss E-1B 148132 (VAW-111, USS Oriskany), operational loss (non-combat) 8 October 1967, 5 crew killed

E-2 Hawkeye—2 total, 0 in combat

-E-2A 151711 (VAW-116, USS Coral Sea), 8 April 1970, 5 crew killed

-E-2B 151719 (VAW-115, USS Midway), 11 June 1971, 5 crew missing

EKA-3 Skywarrior— —2 lost, 0 in combat

-EKA-3B 142400 (VAQ-132, USS America), operational loss (non-combat) 4 July 1970, 3 rescued

-EKA-3B 142634 (VAQ-130, USS Ranger), operational loss (non-combat) 21 January 1973, 3 crew killed

EA-1 Skyraider—4 total, 1 in combat

-First loss EA-1E 139603 (VAW-111, USS Yorktown), operational loss (non-combat) 15 April 1965, crew rescued

-Final loss EA-1F 132543 (VAW-13, USS Franklin D Roosevelt), operational loss (non-combat) 10 September 1966, crew rescued

F-4 Phantom—138 total, 75 in combat

-First loss F-4B 151412 (VA-142, USS Constellation), operational loss (non-combat) 13 November 1964, crew rescued

-Last combat loss (also last USN combat loss of war) F-4J 155768 (VF-143, USS Enterprise), AAA South Vietnam 27 January 1973, CDR H.H. Hall and LCDR P.A. Keintzer POW

-Final loss F-4J 158361 (VF-21, USS Ranger), operational loss (non-combat) 29 January 1973, crew killed

F-8 Crusader—118 total, 57 in combat

-First loss F-8D (VF-111, USS Kitty Hawk), to AAA over Laos 7 June 1964, LCDR C.D. Lynn rescued

-Final loss (operational) F-8J 150887 (VF-191, USS Oriskany) 26 November 1972, pilot rescued

KA-3 Skywarrior- —2 lost, 0 in combat

-KA-3B 142658 (VAH-4, USS Oriskany), operational loss (non-combat) 28 July 1967, 1 crew rescued, 2 killed

-KA-3B 138943 (VAH-10, USS Coral Sea), operational loss (non-combat) 17 February 1969, 3 crew killed

RA-5 Vigilante—27 total, 18 in combat

-First loss RA-5C 149306 (RVAH-5, USS Ranger), operational loss (non-combat) 9 December 1965, 2 crew killed

-Final loss RA-5C 156633 (RVAH-13, USS Enterprise), to MiG-21 North Vietnam 28 December 1972, LCDR A.H. Agnew POW, Lt. M.F. Haifley KIA

RF-8 Crusader—29 total, 19 in combat

-First loss RF-8A (Det. C VFP-63, USS Kitty Hawk), 6 June 1964, to AAA in Laos, Lt. C. F. Klusmann POW

-Final loss RF-8G 144608 (VFP-63, USS Oriskany), operational loss (non-combat) 13 December 1972, pilot rescued

S-2 Tracker—5 total, 3 in combat

-First loss S-2D 149252 (VS-35, USS Hornet), unknown combat loss 21 January 1966, 4 crew MIA

-S-2E 152351 (VS-21, USS Kearsarge), combat loss 11 October 1966, 4 crew KIA

-US-2C 133365 (VC-5, NAS Atsugi, Japan), combat loss 13 May 1967, 2 crew KIA

-US-2C 133371 (VC-5, USS Hornet), operational loss (non-combat) 27 September 1967, crew rescued

-Final loss S-2E (VS-23, USS Yorktown), unknown combat loss 17 March 1968, 4 crew KIA

USN fixed-wing shore-based

  C-47 Skytrain (1)

OV-10 Bronco (7)

P-2 Neptune (4)

P-3 Orion (2)

USN rotary-wing

SH-2/UH-2 Sea Sprite— —12 lost, 0 in combat

-First loss UH-2A 149751 (HC-1, USS Hancock), operational loss (non-combat) 10 January 1966, 4 crew rescued

-Final loss UH-2C 149767 (HC-1, USS Bon Homme Richard), operational loss (non-combat) 10 August 1969, 4 crew rescued

SH-3 Sea King— —20 lost, 8 in combat

-First loss SH-3A 148993 (HS-2, USS Hornet), AAA North Vietnam 13 November 1965, 4 crew rescued

-Final loss SH-3D 156494 (HS-7, USS Saratoga), operational loss (non-combat) 31 December 1972, crew rescued

The United States Marine Corps

US Marine Corps aircraft lost in combat included 193 fixed-wing and 270 rotary wing aircraft

USMC fixed-wing

A-4 Skyhawk—81 lost

A-6 Intruder—25 lost

C-117 Skytrain—2 lost

EA-6A Intruder—2 lost
EF-10 Skynight—5 lost
F-4 Phantom—95 lost, 72 combat
F-8 Crusader—21 lost
KC-130 Hercules—4 lost
O-1 Bird Dog—7 lost
OV-10 Bronco—10 lost
RF-4 Phantom—4 lost
  RF-8 Crusader—1 lost
  TA-4 Skyhawk—10 lost
  TF-9 Cougar—1 lost
USMC rotary-wing

Downed UH-1 Iroquois
AH-1 Cobra – 7
HUS-1 – 75
UH-1E Huey –1969
CH-37 Mojave – 1
CH-46D Sea Knight – 109
CH-53 Sea Stallion – 9

United States Army
USA fixed-wing
OV-1A Mohawk – 3 lost
OV-1B Mohawk – 2 lost
O-1 Bird Dog – 297 lost

More than 67 OV-1afterburner/C/D series aircraft were lost in combat, many of them in Laos and North Vietnam. In just seven months in 1966, the 20th ASTA/131 Aviation Company (Mohawks) lost 6 OV-1As, 20 OV-1Bs, and 2 OV-1C.

USA rotary-wing
5,086 (which include not in addition to the above statistics)
1 Bell 205 (Air America)
270 AH-1G
1 BELL
14 CH-21C
2 CH-34
1 CH-37B
1 CH-37C
83 CH-47A destroyed
20 CH-47B destroyed
29 CH-47C destroyed
9 CH-54A destroyed
3 H-13D destroyed
2 H-37A destroyed
147 OH-13S destroyed
93 OH-23G destroyed
45 OH-58A destroyed
842 OH-6A destroyed
60 UH-1 destroyed
357 UH-1B destroyed

365 UH-1C destroyed
886 UH-1D destroyed
90 UH-1E destroyed
18 UH-1F destroyed
1313 UH-1H destroyed
176 UH-34D destroyed

Glossary

ADVERSE YAW: The tendency of an aircraft to yaw away from the applied aileron. Induced by rolling motion and aileron deflection, usually greatest at a high angle of attack and full aileron deflection.

AILERON ROLL: Rolling the aircraft around the longitudinal axis by use of the ailerons.

AIRCRAFT AXES: There were three axes, which were mutually perpendicular and had a common point of intersection. The longitudinal axis parallel to the fuselage reference line and aircraft rotation around this axis called roll or bank. The aircraft vertical axis was perpendicular to the longitudinal axis through the center of the plane, and rotation about that axis was yaw. The lateral axis was perpendicular to both other axes at the point of intersection. Rotation around that axis was pitch,

ANGLE OF ATTACK (AOA): The angle between the chord line of the wing and the aircraft flight path (relative wind).

ANGLE OFF: The angular measurement between the longitudinal axis of an aerial target and an attacking aircraft measured from the target's tail assuming the attacker was flying a pursuit attack.

ASPECT ANGLE: The angle measured from the tail of an aircraft to define position relative to another aircraft regardless of the relationship of the flight path.

BARREL ROLL ATTACK: A maneuver used at high angles off and long ranges to arrive closer to the lethal envelope.

BARREL ROLL (HIGH G): A maneuver used to cause an overshoot by an attacker who had maintained position in the lethal envelope during maximum performance turning, a last-ditch maneuver leaving the defender with little or no maneuvering potential.

BREAK: A maximum performance defensive turn into the attacker instantly destroying an attacker's tracking solution or position in the defender's lethal envelope.

CLOSURE (Relative Velocity): The time rate of change of distance along the line of sight between aircraft

CONSTANT PEG Air Combat Training with MiG from 4477th TES, at TTR, the early 1980s - 03/1988, follow-on to HAVE IDEA.

DEFENSIVE SPLIT: A controlled separation of a defensive element in different planes used to force the attackers to commit themselves to one of the defenders. A controlled separation with the defenders turning in the same relative direction, however, separated in the horizontal and vertical planes.

DEFENSIVE TURN; The basic defensive maneuver designed to prevent an attacker from achieving a launch or firing position. A planned turn to prevent an attacking aircraft entering the defender's lethal envelope. Defensive turns can be from one g to maximum performance.

DIVING SPIRAL: A near vertical accelerating dive using g and roll rate to destroy an attacker's tracking solution and gain lateral separation.

DOM: Defensive combat maneuvering.

ELEMENT: The basic fighting unit (two aircraft).

ENERGY LEVEL (Es): Total energy state possessed for a given combination of altitude and airspeed (Each).

ENERGY MANEUVERABILITY: A concept used to determine total in-flight performance by measuring instantaneous and sustained maneuverability of an aircraft through its envelope.

ENERGY RATE (ERs) A measure of the ability to gain or lose energy regarding altitude and airspeed or combinations thereof.

FME Foreign Material Exploitation

HARD TURN (Single Direction Turn): A planned defensive turn in which by the angle-off, range, and closure of the attacking aircraft governs the intensity of the turn.

HARMONIZATION: The adjustment of guns and sight of an aircraft, so when within effective range, the tracking index indicates the impact point of the bullets.

HAVE DRILL / HAVE FERRY Exploitation of 2 ex-Syrian MiG-17F from Israel, used for Air Combat Training at Groom Lake, USAF/USN joint project, predating Have Idea (1969)

HAVE DOUGHNUT Exploitation of 1 MiG-21F-13, used for Air Combat Training at Groom Lake, USAF/USN joint project

HAVE GLIB evaluation of foreign radar and threat systems. The systems used names such as Mary, Kay, Susan, and Kathy and arranged to simulate a Soviet-style air defense complex. November 1970

HAVE IDEA Evaluation MiG-21 and MiG-17F variants. May 1973

HAVE PAD Exploitation of the Flogger MiG-23MS (1978)

HIGH-SPEED yo-yo: An offensive maneuver performed to maintain nose-tail separation and prevent the possibility of becoming engaged in a scissors maneuver. A maneuver used by an attacker, in the vertical and horizontal planes, to prevent an overshoot in the plane of the defender's turn.

IN-TRAIL: Individual aircraft, one behind the other.

JINKING MANEUVER: A series of rapid turn reversals or abrupt changes of roll/pitch angle at random intervals, to prevent an attacker from achieving a tracking solution. Usually employed with little load factor while gaining lateral separation.

LAG PURSUIT ATTACK; An attack in which the nose of the attacker's aircraft remains pointed behind the defender's aircraft

LATERAL SEPARATION: Distance between an attacker and defender, measured in the horizontal plane perpendicular to the defender's flight path.

LEAD PURSUIT ATTACK: An attack in which the nose of the attacker's aircraft remains pointed ahead of the defender's aircraft

LETHAL ENVELOPE: An area around every aircraft from which an attacking aircraft can launch a missile or achieve a gun-tracking solution and expect to down the defending aircraft

LOW-SPEED yo-yo: A maneuver employed to facilitate closure and at the same time allow an attacker to remain inside an opponent's turn radius. A maneuver to close on a target using a combination of altitude, airspeed, and cutoff.

LUFBERY: A circular tail chase

MANEUVERABILITY: The ability to change direction and/or magnitude of the velocity vector.

MANEUVERING ENERGY. The ability to perform maneuvers because of energy possessed.

MAXIMUM: Maximum afterburner power.

MAXIMUM MILITARY POWER: Maximum non-afterburner power.

MAXIMUM PERFORMANCE: The best possible performance without exceeding aircraft limitations.

MAXIMUM PERFORMANCE MANEUVERING ENVELOPE: A maneuvering region for the wingman to achieve maximum performance maneuvers with optimum visual coverage and mutual support.

MAXIMUM RATE TURN: That turn at which the plane achieves the maximum number of degrees per second.

RHAW: Radar Homing and Warning

SANDWICH: A maneuver designed to place the attacking aircraft/element(s) in the train between the defending element(s).

SEPARATION: Distance between an attacker and defender. Could be either lateral or longitudinal.

SCISSORS: The execution of a defensive maneuver in which a series of turn reversals attempt to achieve the offensive after an overshoot by the attacker.

SPLIT PLANE MANEUVER: Maneuver involving two aircraft in a defensive split, or four aircraft in a fluid separation to force attacker(s) to commit on one aircraft or element.

TCA: Track Crossing Angle TOA-Angle-Off (Aspect Angle): The angle between the defender's line of flight and the attacker's line of sight measured in degrees (Track Crossing Angle).

YF-110B MiG-21F-13 "Fishbed-C."

YF-110C A MiG-21 "Fishbed" variant

YF-110D A MiG-21 "Fishbed" variant

YF-113A MiG-17F "Fresco-C" used in HAVE DRILL program

YF-113B MiG-23BN "Flogger-F."

YF-113C MiG-17F (a Chinese-built J-5) "Fresco-C" used in HAVE PRIVILEGE program

YF-113E MiG-23MS "Flogger-E."

YF-114C MiG-17F "Fresco-C" (the one used in the HAVE FERRY program)

YF-114D MiG-17PF "Fresco-D

V-N Diagram: A plot of load factor versus velocity used to provide a measure of instantaneous maneuverability.

List of Acronyms and Abbreviations

afterburner   Afterburner

ACM   Air Combat Maneuvering

AERO1A   F-4B Missile Control System

AFFTC   Air Force Flight Test Center

AGL   Above Ground Level

AIM-7E   SPARROW III Missile

AIM-7E-2   SPARROW III (dogfight) Missile

AIM-9B/D   SIDEWINDER Missile

AMCS   Airborne Missile Control System

AN/APR-25   Radar Homing and Warning System

APG-59   F-4J Radar

APQ-72   F-4B Radar

APQ-94   F-8E Radar

ATOLL   Soviet Air-to-Air Infrared Missile

AWG-10   F-4J Missile Control System

Bingo   Minimum fuel state required for safe return to base

B/N   Bombardier/Navigator

CNO   Chief of Naval Operations

COMOPTEVFOR   Commander Operational Test and Evaluation Force

CRT   Combat Rated Thrust

CW   Continuous Wave

DIA   Defense Intelligence Agency

DRV   Democratic Republic of Vietnam

ECM   Electronic Counter Measures

ECCM   Electronic Counter Counter Measures

EGT   Exhaust Gas Temperature

EI   ELINT - Electronic Intelligence

FFAR   Folding Fin Aircraft Rocket

FME   Foreign Material Exploitation

Free F-4   Tactical Wingman

FTD   Foreign Technology Division of Air Force Systems Command

g   Acceleration due to gravity

HAVE DOUGHNUT   Project Name

HEI   High Explosive Incendiary

I Band   Radio Frequency Energy (8,000 to 10,000 KC)

IMN   Indicated Mach Number

IR   Infrared Radiation

KCAS   Knots Calibrated Airspeed

KIAS   Knots Indicated Airspeed

KTAS   Knots True Airspeed

MBC   Main Beam Clutter
MiG-21   MiG-21 F-13 (Fishbed C/E)
MM   Millimeter
MRT   Military Rated Thrust
MSL   Mean Sea Level
NATC   Naval Air Test Center
NATOPS   Naval Aviation Training and Operational Standardization Program
NM   Nautical Miles
Oblique Loop   An overhead maneuver performed just off the true vertical. An angle of bank held to facilitate maintaining visual contact with a target in the rear hemisphere
Padlock   A lookout technique that required looking solely at the visually acquired target.
PLM   Pilot Lock-On Modification
PPS   Pulses Per Second
PRF   Pulse Repetition Frequency
PD   Pulse Doppler
Pk   Predicted Kill
q   Dynamic Pressure
SEA   Southeast Asia
TAC   Tactical Air Command
TAC LEAD   Tactical Lead
TAC WING   Tactical Wing
TCA   Track Crossing Angle
UHF   Ultra High Frequency (Radio)
VA   Fixed-wing, heavier than air, attack aircraft
Vc   Closing Velocity
VF   Fixed-wing, heavier than air, fighter aircraft
VID   Visual   Identification

References

Foreign Technology Division, US Air Force Systems Command FTD-CS-09–5–67, Secret Publication of 18 July 1968, "Fresco (MiG-17) Weapon System "

NAVAIR-01–245FDB-1 "NATOPS Flight Manual F-4B Aircraft" of 15 February 1969

NAVAIR 01–245FDB-1A "Supplement to NATOPS Flight Manual F-4B Aircraft" of 15 February 1969

NAVAIR 01–245FDD-1 "NATOPS Flight Manual F-4J Aircraft" of 15 Jun 1967

NAVAIR 01–245FDD-1A "Supplement to NATOPS Flight Manual F-4J Aircraft" of 1 Jun 196 8

NAVAIR "01–45HHE-1 NATOPS Flight Manual F-8H Aircraft" of 1 Could 1967

NAVAIR 01–45HHE-1A "Supplement to NATOPS flight Manual F-8H Aircraft" of 1 Could 1967

NAVAIR 01–45HHF-1 "NATOPS Flight Manual F-8J Aircraft" of 15 April 1968

NAVAIR 01–45HHF-1A "Supplement to NATOPS Flight Manual F-8J Aircraft" of 15 April 1968

NAVAIR 01–40afterburnerE-1, "NATOPS Flight Manual A-4/Ta-4 " of 15 February 1967

NAVAIR 01–85ADA-1, "NATOPS Flight Manual A-6A " of 1 December 1967

NAVAIR 01–45AAA-1 "NATOPS Flight Manual A-7 " or 1 January1968

NAVAIR 01–245FDB-1T Tactical Manual for F-4B/J Aircraft of 1 March 1969

NAVAIR 01–245FDB-1T Supplement to Tactical Manual F-4B/J Aircraft of 1 March 1969

NAVAIR 01–45HHA-1T Tactical Manual for F-8 Aircraft of 15 September 1968

NAVAIR 01–45HHA-1T Supplement to Tactical Manual for F-8 Aircraft of 1 September 1968

NAVAIR 01–40AV-1T, "A-4/TA-4 Tactical Manual " of 1 April 1967

NAVAIR 01–85ADA-1T, "A-6 Tactical Manual " of 15 December 1967

NAVAIR 01–45AAA-1T, "A-7 Tactical Manual " of 15 April 1968

Weapons Systems Evaluation Group (WSEG) Secret Report #116, of 20 October 1967, "Account of F-4 and F-8 Events before March 1967 "

National Security Archive October 29, 2013

http://www2.gwu.edu/~nsarchiv/NSAEBB/NSAEBB443/

http://www2.gwu.edu/~nsarchiv/NSAEBB/NSAEBB443/docs/area51_48.PDF

http://www2.gwu.edu/-nsarchiv/NSAEBB/NSAEBB443/docs/area51_49.PDF

http://www2.gwu.edu/~nsarchiv/NSAEBB/NSAEBB443/docs/area51_50.PDF

RED EAGLES, America's Secret MiG by Steve Davies

America's SECRET MiG Squadron by Gail Peck

Campbell, John M., and Hill, Michael. Roll Call: Thud. Atglen, PA: Schiffer Publishing Ltd., 1996. ISBN 0–7643–0062–8.

Hobson, Chris. Vietnam Air Losses, USAF, USN, USMC, Fixed-Wing Aircraft Losses in Southeast Asia 1961–1973. North Branch, Minnesota: Specialty Press, 2001. ISBN 1–85780–115–6.

FTD-CR-20–13–69-INT Vol I 3 October 1997

FTD-CR-20–13–69-INT Vol II August 23, 2000

FTD-CR-20–02–69 Vol I April 1970 Declassified October 3, 1997

FTD-CR-20–02–69 Vol II April 1970 Declassified 3 October 1997

Mayday by Beschloss, Michael R., Harper & Row 1986

Memories of Dr. Franklin Allison Rodgers, Ph.D. February 3, 1995

Thornton D. "TD" Barnes, author, and entrepreneur, grew up on a ranch at Dalhart, Texas. He graduated from Mountain View High School in Oklahoma and embarked on a ten-year military career. He served as an Army intelligence specialist in Korea and then continued his education while in the US Army, attending two and a half years of missile and radar electronics by day and college courses at night. Barnes deployed with the first combat Hawk missile battalion during the Soviet Iron Curtain threat before attending the Artillery Officer Candidate School, where an injury ended his military career.

Barnes's career includes serving as a field engineer at the NASA High Range in Nevada for the X-15, XB-70, lifting bodies and lunar landing vehicles; working on the NERVA project at Jackass Flats, Nevada; and serving in Special Projects at Area 51. Barnes later formed a family oil and gas exploration company, drilling, and producing oil and gas and mining uranium and gold.

Barnes currently serves as the CEO of Startel, Inc., a landowner, and is actively mining landscape rock and gold in Nevada. He serves as the president of Roadrunners Internationale, an association of Area 51 veterans, and is the executive director of the Nevada Aerospace Hall of Fame.

Two National Geographic Channel documentaries feature Barnes: Area 51 Declassified and CIA—Secrets of Area 51. Numerous documentaries on the History Channel, the Discovery Channel, the Travel Channel, and others also feature him. Barnes is the author on several books approved by the CIA PRB including the CIA Area 51

Chronicles, a three-book series about the CIA at Area 51. Barnes lives in Henderson, Nevada.

Connect with the Author Online
Facebook: www.facebook.com/ThorntondBarnes
Website: td-barnes.com
LinkedIn: www. LinkedIn.com/profile/edit?trk=tab_pro
Twitter: twitter.com/ThorntonDBarnes

Other Books by the Author

Fiction
EMP - Book 1 - Nuclear Winter
EMP - Book 2 - Nuclear Spring
EMP - Book 3 - Nuclear Summer
The Wildcatter
The Senator
The White Hats
Non-fiction
Soaring with the Eagles
The Secret Genesis of Area 51
THE AREA 51 CHRONICLES the CIA AT AREA 51 1955–1979
    Book 1 - The Angels
    Book 2 - The Archangels
    Book 3 - The Company Business

Books available on CreateSpace, Smashwords, Apple, Amazon, Kobo, Barnes & Noble and Sony.

eBooks are available at Smashwords, Nook, iTunes, KOBO, and Amazon with distribution in the USA, UK Germany, France, Spain, Italy, and Japan.

Available in paperback at Createspace, Barnes & Noble, and Amazon with distribution in the USA, UK, Germany, France, Spain, Italy, Netherlands, Japan, Mexico, Australia, India, Brazil, and Canada.

www.ingramcontent.com/pod-product-compliance
Lightning Source LLC
Chambersburg PA
CBHW080009210526
45170CB00015B/1957